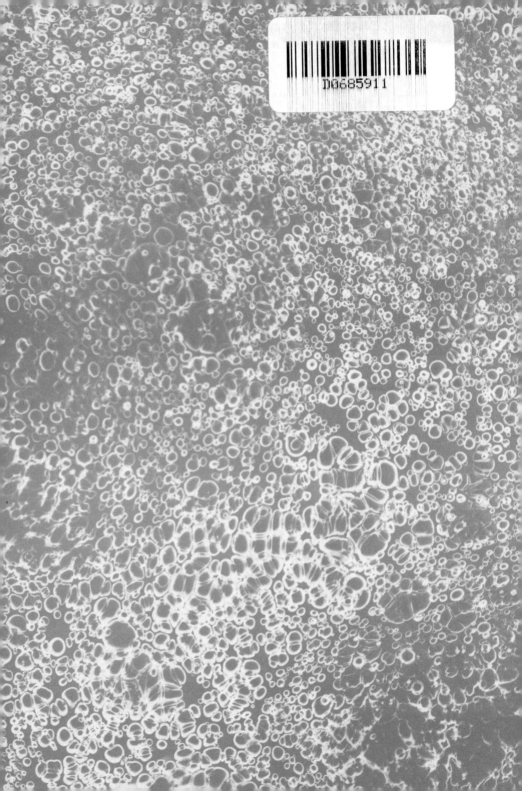

Blyth in the Eighteenth Century

From an original in the possession of Blyth Corporation

Painted about 1827 by Balmer

BLYTH
IN THE
EIGHTEENTH CENTURY

by

W. R. Sullivan

ORIEL PRESS

ISBN 0 85362 130 6
Library of Congress Catalogue Card No 70–166003

Publication of this book has been assisted by
The Northumberland County Council.

Published by Oriel Press Limited
32, Ridley Place,
Newcastle upon Tyne, England NE1 8LH.

Text Set in Times Roman
Printed by R. Ward & Sons, Limited, Newcastle upon Tyne.

Preface

THE NORTHUMBERLAND COUNTY RECORD OFFICE contains a rich collection of source material on the county's history, and the object of the present study is to use a small part of the collection to illustrate the history of Blyth in the eighteenth century. Some material from parish registers and other sources is also included, but the main body of evidence is drawn from the Ridley Papers deposited in the Record Office by Lord Ridley, and thereby made available to local historians interested in the history of the county, in which the Ridley family over many generations played a notable role.

The general plan of the work is as follows: after a short introduction there is a note on the various sources used and their nature. Then we have brief introductions to a series of sections illustrating different aspects of life in eighteenth century Blyth. The main body of the work consists of a selection of documents and other original material; these are numbered and bracketed references in the introductory sections refer to the documents concerned. So far as possible the documents have been faithfully transcribed as they appear in the original but some minor changes in layout have been made. The identity of each document is given in the list of references on page 94, in which any omissions or alterations are also noted.

Acknowledgments

THE AUTHOR gratefully acknowledges the co-operation of Lord Ridley, whose family papers formed the background of this work. The work would not have been possible without the constant help and advice of Mr. Robin Gard, County Archivist of Northumberland and the whole of his staff; while the assistance of Dr. Norman McCord of the University of Newcastle upon Tyne in revising the text was invaluable.

The staff of Newcastle upon Tyne Central Library gave much help in the examination of the files of the *Newcastle Journal* and other records, and the Customs and Excise staff at all the relevant ports showed that the routine of official duties had not dulled their appetite for local history.

Finally, since in this day and age costs make the publication of a book of this nature impossible without help, the assistance of the County Council is gratefully acknowledged.

Contents

1 Introduction

BLYTH is a port on the Northumberland coast eight miles north of the River Tyne and fourteen miles distant from Newcastle upon Tyne by road. In the nineteenth and early twentieth centuries Blyth was an important centre of coal mining, coal shipping, shipbuilding and ship repairing, and during that period its development was rapid. Since then the character of the town's economy has markedly altered. Only one pit continues at work, and an important power station at Cambois draws fuel from a wider area of the Northumberland coalfield. The shipbuilding industry ceased for some years in the 1960s and although it has recently re-commenced on a reduced scale Blyth has been seriously affected with other parts of the North East by the closure and run-down of its once staple industries. Efforts to fill the economic vacuum and to provide a more varied pattern of employment have had some success through the development of a trading estate, while the building of new houses has made Blyth into an important residential area. In the eighteenth century, however, Blyth was a very different place, and the evidence offered here is intended to illustrate the basis upon which the mushroom growth of Victorian Blyth was erected.

Although this material may seem almost parochial, it is not unimportant. The development of the industrial revolution in Britain produced the world's first industrial society, and provided the basis for the empire of Victorian Britain, one of the greatest empires the world has ever seen, with influence extending into every quarter of the globe. In the growth of imperial Britain places like Blyth were to play a far from negligible role. A significant part of the energy sources of Victorian Britain, for instance, flowed from the Great Northern Coalfield through shipping ports like Blyth. Thus, in this study the evidence for the development of the town in the earlier phases of the industrial revolution sets the scene for the massive expansion which was to follow. Already eighteenth century Blyth possessed its infant industries; if the scale was small, the technology often primitive, and the methods of administration and control somewhat haphazard and experimental, yet these were the growing pains of a new industrial society which in its eventual flowering was to be part of some

1

of the most important changes in the nature of human society through-out history.

The eighteenth century saw many moves towards technological inno-vation, and the history of Blyth in that period shows some of them at work, even though here there was nothing quite as impressive as the developments later in the century in the textile areas of Britain.

The area with which we are concerned is smaller than the later borough of Blyth. It is confined to a tract of land stretching from Meggie's Burn in the south, to the line of the Plessey waggon-way in the north, as far as the slake and thence by that water channel to the river (the Plessey waggon-way followed exactly the line of the present Plessey Road and the Slake that of Union Street), with a small extension of some seven acres to the north of the waggon-way in the vicinity of the old grammar school. The River Blyth and the sea bound our area on the east, and the western boundary is the limit of the old parliamentary constituency, running a little to the west of the present Newcastle Road, Newsham. The total area with which we are concerned is 1,178 acres, and the first census of 1801 showed a population of 1,170—the scale of subsequent development can be seen by the latest estimate of Blyth's population at 35,130.

The relevant area belonged at the beginning of the eighteenth century to the Radcliffe family, Earls of Derwentwater, as owners of the Lord-ship of Newsham. Their tenure began only in 1694, when they had bought the lordship for £13,000, and was not to be long-lived. The third and last Radcliffe, Earl of Derwentwater, was one of the peers executed after the 1715 Jacobite rising, and his property and that of his family was confiscated.

In 1723 the Lordship of Newsham was bought jointly by Matthew White of Newcastle and his son-in-law Richard Ridley of Heaton, both men with important and growing interests in the City of Newcastle. The transaction is of considerable interest in social history as an example of the way in which it was possible for men whose fortunes had been accumulated in town-based trade to acquire estates and move into the country society. In 1730 Matthew White extended his landed possessions by the purchase of Blagdon which still remains the seat of his descend-ants. The rise of the Ridley family from merchants into the landed gentry and then into the aristocracy is a very good example of that kind of social transition. Similarly, the mixed economic interests which these men thereby acquired represents a characteristic feature of the economic development of the North East region, where it was common for the

2

landowning class to participate effectively in industry and commerce. Certainly, Matthew White and Richard Ridley acquired their interest in Blyth in the early eighteenth century when they were already successful in business and well aware of potential sources of wealth. They were determined to make their new property a paying concern.

Link House Farm which was within the area of the Lordship of Newsham appears to have been in different ownership to the rest of the area. Matthew Robinson of Link House left his estate to Matthew Ridley of Heaton, in trust for Richard Ridley by will dated 16th July, 1748. This Richard Ridley, born in 1735, was Matthew Ridley's son by his first wife Hannah Barnes. In the survey of 1785 Link House Farm is shown as belonging to Nicholas Ridley. Nicholas would appear to have been the younger brother of the second baronet and later when he became a Baron he took the name of Colborne after his mother Sarah Colborne.

The purchase of Newsham was supplemented on the first day of 1752 when the then Matthew Ridley acquired rights in the River Blyth previously possessed by the Bishop of Durham (document I). It seems that as the port developed these rights became much more valuable than the ten shillings rental attributed to them in 1752. The conveyance of 1752 did not finally settle the question of the Bishop's rights, since after differing interpretations of the agreement had been advanced in 1786 the dispute lingered on until 1821 when a compromise settlement was at last agreed.

The value of the Blyth property acquired by the Ridley family can be gathered from two rent rolls of 1698 and a survey of the Lordship, of Newsham in 1785 (2 and 3). In both of the rent rolls the farms and their tenants are named. The 1785 survey is a fuller document, giving farms, tenants, and acreages (in acres, roods and perches); the fields are numbered, and accompanying maps use the same numbers (41). In 1785 none of the seven farms on the estate exceeded 195 acres, the smallest being 114 acres and the average size 146 acres. Some evidence of the area under tillage comes from a document of 1787 (4), which gives the grain acreage as 358. Details of crops grown in the years 1812 and 1815 are given on the back of a map of Link House Farm (5), which conveniently uses the same field numbers as the survey of 1785. Only one of the old field names used in these old maps and accounts is still in use today, a field called "Shoulder of Mutton." Almost all the field boundaries,

3

however, on the other hand have for the most part survived to the present day. The whole area is shown in a map of 1840 in Section 13.

The surviving Ridley Papers provide no record of cattle holding, but accounts in the Newcastle Journal of 1750 and 1751 of outbreaks of cattle disease, described as distemper locally but verified as rinderpest, mention three hundred cattle in the Blyth District. These accounts also show that an energetic policy of slaughtering and burying the infected beasts was introduced under regulations laid down by the government, with payments of subsidies for mature beasts and calves.

The correspondence and accounts for the Ridley estate at Blyth show clearly that for most of the eighteenth century at least the landed property, whatever social prestige it may have carried, remained less important in the family's income than a variety of business and banking activities. However, Blyth was expected to pay its way and the extent to which it did so is illustrated by a summary account covering the years 1789-99 (6). Even so, a letter dated 14th June, 1776 (25) shows that money-making was not the only consideration.

The spirit in which the Ridley family approached their Blyth property can be well illustrated by the following advertisement which appeared in the *Newcastle Journal* on 17th January, 1744.

"At Blyth a good seaport in Northumberland, good convenience for carrying on any trade, with liberty to build warehouses, granaries and other things necessary; also a new windmill built with stone and well accustomed, a fire-stone quarry for glasshouse furnaces, a draw-kiln for limestones, two large sheds for making pan-tiles and stock bricks, with a good seam of clay for that purpose.

Also at Link House, one mile from Blyth, a large new malting well supplied with water. Enquire at Link House aforesaid or of Matthew Ridley Esq., Newcastle".

2 *The Coal Trade*

COLLIERY accounts year by year provide the most substantial item in the financial records of the Ridley holdings at Blyth, despite the fact that the pits were in fact some distance from the port. The mines were shallow workings in the area of Plessey Checks and Bassington farm. This was an area of extensive workings; a plan in volume 9 of the Northumberland County History identifies eighteen pits there, although the Ridley accounts for our period mentions only six by name; the West Pit, South Pit, Rodney Pit, Rising Sun Pit, Hall Pit and Success Pit. The production of these pits for the years 1783-88 shows that these mining ventures were a profitable enterprise (7). The amounts were given in "tens", a unit of weight used in Northumberland and Durham, varying according to individual agreements between pitowner and royalty owner, at anything between 48 and 50 tons. Most of the coal from these pits was distributed by sea from Blyth, but there was a useful local sale too, as a schedule of disposal for 1788 shows (8).

The reason why the pit accounts appear in the Blyth documents was presumably because most of the coal was conveyed to Blyth by a waggon way from Plessey Checks. Its route went by Bog Houses and New Delaval and then along Plessey Road to the shipping point. Part of the line of this early railway can still be seen as a double hedge running alongside the modern road running east from Plessey Checks and as another double hedge on the brow of the hill east of Bog Houses. (This double hedge has now disappeared in a road improvement).

The maintenance of the waggon-way was a heavy charge on the estate, amounting to £1,140 in 1783, £843 in 1784, £214 in the first quarter of 1785, and £830 in 1786. The wayleave rent of £300 to Lord Delaval was the largest recurring item but the repairs to the "line" itself were costly, oak and ash being mostly used for both rails and sleepers. In the 1786 and 1787 accounts there are exceptional and unexplained entries for £42. 10. 7. and £22. 8. 3. for rails and wheels supplied to Bedlington Colliery. The land tax on the waggon-way was £7. 5. 0. in 1786 while in the following year a half year's tax is entered as £1. 16. 3. The account for 1784 copied here (9) is typical, although since there is no entry for snow clearing, amounting to £2. 2. 10. in 1783, this suggests that the winter was an unusually mild one.

5

The line ran to a shipping point on the River Blyth. As early as 1772-73 there are references to loading coal there from some kind of "spout". In 1784 we have a detailed account (11) of expenditure on a "staith". The meaning of this word at that time is not clear, for it was used either for a depot, a landing, or a wharf from which coal could be loaded. The account for 1784 includes items which seem likely to relate to construction work rather than the cost of regular working, and references in contemporary letters rather suggest that this Blyth "staith" should be interpreted as a coal storage depot rather than any technical advance in actual loading methods.

Until late in the nineteenth century the condition of the Blyth harbour presented many navigational hazards, as a report of 1814 from the distinguished engineer John Rennie illustrates (12). Nevertheless the port was increasingly busy. Records of colliers using the port and their cargoes survive for the years 1755-67 and 1793-99 (13). These documents illustrate predictable seasonal variations, as well as changes in the ports of registration from which ships came to Blyth, and a discernible increase in the average cargo carried. (13, 14, 15, 16). A comparable source for the very early nineteenth century is included in a large book of uncertain provenance, which was found floating in the harbour and is now in Blyth Borough Public Library. This gives a good deal of detail of coal sales for 1803-04.

The developing coal trade of the later eighteenth century produced its own crop of problems and difficulties, and the Ridley Papers contain a large number of letters which passed between the Ridley family and their agents on such matters. The letters from Matthew White Ridley, to his London agent, George Ward, during the years 1767 -77 are particularly full. A few examples of these are reproduced here (17). Even in those days of poor communications Ridley was able regularly to reply to letters on the fourth day after they had been written in London. One matter which more concerned northern coal-owners was the duties on coal imposed in London, which were both complex and vexatious. (18, 19). It is not surprising that historians of this period have found the varying weights and measures difficult to interpret for they were just as confusing to people at the time as one paper reveals (20).

By the end of the eighteenth century the shipment of coal from Blyth, if still much smaller than its Tyne neighbour, was firmly set on an upward course which was to continue for many years to come (21).

3 Salt-making

SALT-MAKING has a history in North East England going back to med-
ieval times. In 1825 the local historian Eneas Mackenzie wrote that in 1605
the two counties of Northumberland and Durham possessed 153 salt pans
employing 430 salters as well as many ancillary workers. Blyth by the early
eighteenth century had its own share in this trade. The Ridley Papers
contain documents, for example, of 1723, 1725 and 1731 relating to these
works. Wallace, in his early history of Blyth, writes of eight salt pans at
Blyth itself, four at North Blyth and two where the Folly now stands. The
eight pans at Blyth are recorded in an estate map in the Ridley Papers, at
a point about 50 yards from the river about midway between the end of the
Plessey waggon-way and the slake. The two "Folly" pans were just west of
the corner of the present Park Road and Plessey Road.

Earlier salt pans were of lead, some of them being quite small; one of
2 feet diameter and 3 inches depth is recorded for Blyth. The size gradually
grew, lead pans of diameter of 5 feet 6 inches being mentioned. Eventually
iron replaced lead, and some of the iron salt pans of the late eighteenth
century were of substantial construction. When one of the North Blyth
pans was broken up in 1744 more than four tons of iron were recovered (23).

It is clear that salt-making was a source of female employment. The only
three entries for salters in the Earsdon Parish register of burials refer to
women, and the Ridley Papers contain a document recording payment of
money to Mrs. Weatherhead for salt-making, Salt-making is well repre-
sented in the surviving detailed accounts (22, 23, 24). One item calling for
explanation is a payment for blood. Both blood and urine were employed
in the salt manufacturing process in these early periods, for the purpose
of precipitating mineral impurities in the boiling liquid.

The salt trade was burdened with taxation for almost all of the eighteenth
century, and in this case the rate of tax steeply increased during the century.
For example, the levy was 5 shillings a bushel of 8 gallons after 1775, and
in 1798 war-time taxation doubled this rate. It was not until 1825 that this
tax was removed, but this was too late to do much to help the coastal
salt-making industry, then in very marked decline. Although the last of the
old salt pans at Blyth was not finally destroyed until 1876, the local industry

seems to have died out in the early nineteenth century. An effort by James Nelson and a man called Douglas to revive the Blyth salt trade about 1838 was a failure. Again, letters in the Ridley Papers illustrate the nature of the industry and its problems in the eighteenth century (25).

4 *The Brewery*

IN THE early eighteenth century there was a brewery at Blyth, supplying both the small local population and the ships using Blyth harbour. A lease of 1725 shows that this undertaking was a small scale enterprise, which seems to have functioned as a publican brewer. An account of 1774 gives some information about quantities of materials involved in its working, and their cost. As Blyth grew, the demand increased and a much larger brewery was erected between 1784 and 1786. Detailed accounts survive of the cost of the construction work (26, 27). The old brewery was situated at the back of Queens Lane and the new one at the side and rear of the Brewery Bar (Pilot Cutter). From the evidence it is quite clear that the new brewery with its greatly enlarged capacity was needed to meet an expanding public demand. The increased scale of production can be illustrated by a comparison of the sale of spent grains by the old and new breweries. In 1774 the old brewery sold 3,021 bushels whereas in 1797 its successor disposed of 6,129 bushels. The sale of beer likewise increased, from an output of 2,806 half bushels in 1789 to 5,318 in 1797.

The lease of the new brewery to Henry Ridley dated 20 January 1798 mentions 23 public houses, twelve of them in Blyth, which were tied to the brewery for purchase of beer. It is noticeable that Sir Matthew Ridley himself took an interest in the brewery; for nine and a half years he operated it directly rather than leasing it, during which period his brewery profits averaged about £400 per annum. Ridley also supplied malt and hops for the brewery's working (28, 29). It is not surprising that already in the eighteenth century drink was a favourite target of taxation, as a list of contemporary duties on beer and its manufacturing materials readily shows (30).

The brewery at Blyth was one of the many smaller breweries to disappear in the changing conditions of the twentieth century. Wartime restrictions forced its closure in 1916 and it never reopened. The plant was disposed of in the late 1930's and the building partly demolished.

9

5 Rope-making

IT WAS natural that in a port like Blyth rope-making should be a local industry, and we know that this manufacture was carried on at least from 1762. The evidence however is not so full as for other enterprises, for the Ridley family do not appear to have actively participated in this industry, and their papers only supply leases of land for ropeworks. These leases show the peculiar nature of these works.

The yarn was spun by hand, the hemp being first hackled by combing it straight over a board studded with sharp steel teeth. A bunch or head of hackled hemp was placed around the spinner's waist, and he then attached a few fibres to a hook on the spinning wheel. The wheel was revolved by hand as he walked backwards feeding the fibre from the supply around his waist. Two or more yarns thus produced made a strand, while three or more strands twisted together made a rope, and three or more ropes were combined to form a cable. Horse power was used where the rope was too large to be made by hand.

All the leases of land were for an area 400 yards long and 6 yards wide, on which the long low buildings known as rope-walks, from the walking to-and-fro of the workers, were erected. Two rope-walks founded in Blyth in 1762 had a long life; they were both still shown on the 1859 Ordnance Survey Map, one on the line of the present Ridley Avenue, the other on the line of Park View. The Ridley Avenue walk was occupied by John Clark and the Park View one by George Marshall, who also occupied a raff-yard and built the dwelling-house which afterwards became the Ridley Arms public house.

Rope-makers were usually associated with sail-makers in sea ports. Details of the leases are shown in (31).

6 Brick-making

IN THE 1785 survey (3) field number 57 of 5 acres is given the name Clayhole; it was in this field, behind the present Wensleydale Terrace, that Blyth's early brickworks were situated. Map No. 16 of Link Farm shows a long narrow building in the centre of the field.

In 1774 this enterprise made 213,600 bricks (32), and in 1788 it appears to have used 43 chaldrons or 114 tons of ching coals (8).

The new brewery was built with Blyth bricks and they were also used in local houses and salt pans, in the Plessey mines and at Blagdon itself.

The Earsdon Parish burial records show three Blyth brickmakers, two of them members of the Turner family. Joseph Turner who was making bricks in 1774 died at Blyth on 18th December 1796.

7 Ship-building

SHIP-BUILDING was to play an important part in Blyth's history, although, again, one in which the Ridley family played no direct part. However, other sources enable us to reconstruct something of the history of local ship-building in the eighteenth century.

The industry seems to have been founded by Edmund Hannay, born in 1727 in Bothal, near Ashington, where his grandfather and father had been parish clerk. In the early 1740's Edmund was in Scotland, living with his uncle Captain William Hannay at Kingsmuir, near Crail in Fifeshire, while learning the ship-building business at Leith. He returned to Northumberland after the '45 Rising and settled at Blyth in 1750, marrying a local girl. They had eight sons and three daughters, but only two daughters survived their father. Edward Watts, who married Hannay's eldest, also started a shipyard in the same vicinity and his son Edmund Hannay Watts carried on the two yards after Hannay's death in 1800, and his own father's death a few years later. Hannay was only 23 years old when he started his yard which was near Nelson Place. His ships had a very high reputation for design and reliability.

Later in the century the names of M. Watson and R. Stoker begin to appear in the records as shipbuilders at Blyth.

To discover Blyth's ship-building record in the eighteenth century recourse must be had to the official shipping registers of the Customs and Excise Department. The compulsory registration of ships commenced only in 1786, though Lloyds had published a list from an earlier date. Lloyds list however is incomplete as regards the place of building of many of the smaller vessels. After 1786 each port of registration entered the details of ships built and registered in its area in a bound book, and sent copies to the central registry at the Custom House in London. This duplication was fortunate for the central records were destroyed by fire in 1818 and only the port records survive for the earlier years.

Ships built at Blyth were usually registered at Newcastle, the nearest port of registry. Between 1786 and the end of 1799 the Newcastle registers list 35 Blyth-built ships, and the recorded details (33) give a good picture of the ships built at Blyth in those years.

Of these 35 ships, 33 were square sterned and only two were pinks, or vessels with very narrow sterns. Twelve were sloops, or single masted vessels with a fore-and-aft rig and a boom mainsail. Sixteen were brigantines, two-masted vessels, square-rigged on the fore-mast and fore-and-aft rigged on the main-mast. Five were snows, vessels similar to a brig with two square-rigged masts and a boom mainsail on a trysail mast stepped immediately aft of the main-mast.

The registers for Kings Lynn which appears to have had more trade with Blyth than any other port in the eighteenth century unfortunately cannot now be found and the registers at Hull, Sunderland and Bridlington produced no additions, but Whitby added three more sloops, and the Lloyds 1786 unofficial list at Newcastle added three brigs. Among the Whitby items is the sloop "Constant Ann" recorded as built in 1740, this boat is understood to have been Hannay's first ship and if so there must be an error; it is possible to begin ship-building at 23 but 13 is not possible!

Of the Blyth-built ships where the builder's name is recorded, Hannay is credited with nine, Watts with eight, and Watson and Stoker with two each. Further Blyth shipbuilders' names begin to appear early in the nineteenth century. The numbers of ships identified as built in Blyth to 1799 (34) shows obvious gaps and we conclude that, unless some new source of information becomes available, the details of all eighteen century Blyth-built ships will never be known.

Of the Blyth ships found in the records, the two barques were of 254 and 201 tons, the brigs and brigantines averaged 164 tons, and the sloops 49 tons. It seems that only a minority of the ships employed in the Blyth coal trade were built in the port (35).

The records contain several interesting cases where ships were altered to increase their size or alter their rig. The brig "Nancy" built in 1784, tonnage 85, length 59 ft. 6 ins., beam 19 ft. 1ins., was re-registered in 1796 as, tonnage 134, length 75 ft. 6 ins. beam 20 ft. 7 ins. and the sloop "Ruby" built at Blyth in 1797 with one mast, and a raised quarter-deck and forecastle, tonnage 48, length 46 ft. 3 ins., beam 16 ft. 5 ins., became in 1818 a flush decked schooner of 70 tons, length 58 ft., beam 17 ft. 1 in. with two masts.

As with coal-mining, in ship-building Blyth ended the eighteenth century with the foundations laid for a much greater expansion in the next century. If the eighteenth century's scale was small, a reputation for Blyth ships had already been established, and four shipyards were already busy when the century opened.

8 *The South Blyth houses*

The extant evidence suggests that in the later eighteenth century the total number of houses on the Ridley properties in Blyth was about 187, including five farm houses, only one of which now survives. The Ridley Papers include an account of rent due on houses as at 11th November 1787 (36). Of these payments 36 were shown as bad debts.

Many leases survive showing that house property was of varying quality and value and that it included shops as well as dwellings. There are leases to two butchers, a baker, a weaver, a ropemaker, a grocer, one of a "general shop" and another referring to a great stable for the waggon way horses.

Brief details of the leases are given (37) and the practice of granting leases for the term of three lives rather than for a limited number of years is illustrated. The conditions of certain leases clearly show the general concern to make the estate profitable. In 1800 Sir Matthew Ridley leased a butcher's shop to George Lough but prudently provided that, if at any time part of the property should be licensed as a public house, then the occupant must purchase all the ale and porter from the brewery owned by Sir Matthew White Ridley.

All in all the picture emerges of a small but growing community living in houses near their places of employment, and enjoying a slowly expanding variety of services.

9 Some facets of the community

THE MOTTO of the Borough of Blyth, incorporated in 1922, is "We grow by industry", and already in the eighteenth century the truth of this motto was apparent. In the early economy of Blyth we can still see the older tradition of employment in which a young man would serve his apprenticeship to a trade under a master craftsman, then work with him as a and eventually become an independent master himself. When the new Blyth brewery was built in the 1780's, it is clear that the carpenter, mason and smith were local craftsmen working on a contract basis, drawing part of their reward from time to time during the period of building, and receiving the balance when the work was completed. The labour force used in such relatively large constructions seems to have been a local one. Outsiders were brought in to instal the new brewing machinery, but otherwise the workmen bore names familiar in local documents.

The existence of such self-contained communities was no doubt caused by the poor road communications. There appears to be no mention of roads in the Ridley Papers, but Armstrong's map of Northumberland in 1769 shows three roads entering Blyth: one from Seaton Sluice, one from Cowpen, and the third following the line of the Plessey waggon way. The first named would appear to have been merely a rough track if the evidence of letters written to Lord Delaval by his agent at Seaton Sluice is accepted (38).

Apart from documents among the Ridley papers, the parish registers of Earsdon Parish, of which the Lordship of Newsham was a part throughout the eighteenth century, give some insight into the contemporary people of Blyth (39. 40). The lists of occupations recorded help to provide a picture of the community, which in its own way was something of a microcosm of the Britain of the day. Outside the shipping community of the little port, with its waggon-way, staith, ropeworks, shipyards, brewery and salt-pans the area was predominantly agricultural.

Britain as a whole was much like this at the end of the eighteenth century a society still predominantly rural but with important areas of industry and mining which in the next century were to expand so vastly and rapidly as to relegate agriculture to second place in the national economy. Few people in Britain in 1800 could imagine such a development: the Blyth of 1900 would have appeared incredible to the inhabitants of the little port in 1800, but especially in coal and in shipbuilding the eighteenth century had sown seeds which were to produce a very remarkable harvest in the next century.

10 *List of Documents and Sources*

1. Lease of rights in the river Blyth, 1752.
2. Two rent rolls of Newsham Estate, 1698.
3. Survey of the Lordship of Newsham, 1785.
4. Tillage of Newsham Estate, 1787.
5. Cropping of Link House Farm, 1812-15.
6. Profits from Blyth and Newsham estates and trade, 1788-99.
7. Coals wrought from Bassington Colliery pits, 1783-88.
8. Coals led to Blyth Staith and their disposal, 1788.
9. Cost of maintenance of Blyth, or Plessey, Waggon Way, 1784.
10. Payments for the maintenance of the Waggon Way, 1774.
11. Cost of maintenance of Blyth Staith, with expenses for Office and keels, 1784.
12. Description of Blyth Harbour by Sir John Rennie, 1814.
13. Monthly numbers of ships loaded at Blyth, 1755-67 and 1793-99.
14. Registration of ships taking coal cargoes, 1765 and 1797.
15. Coal cargoes carried by Blyth ships, 1754-62 and 1790-98
16. Coals shipped and harbour dues at Blyth, 1783-88.
17. Letters from Matthew Ridley relating to the coal trade 1767-75
18. Duties on coals sent to London, (about 1803).
19. Extract from the case of the Northern Coal Owners for the reduction on duties on coal, 1824.
20. Memorandum on weights of Newcastle chaldron and canal ton, about 1822.
21. Coal exports from Newcastle and Blyth, 1798-1818.
22. Accounts for Blyth Salt-Pans, 1783-84, 1790/91.
23. Payments for Blyth Salt-Pans, 1774.
24. Analysis of Blyth salt prices, sales and profits, 1783-99.
25. Letters from Matthew Ridley relating to the Salt Trade, 1767-76.
26. Cost of building the New Brewery, 1784-86.
27. Analysis of costs of building the New Brewery, 1784-86.
28. Accounts for grains, malt and hops, 1774.
29. Blyth Brewery accounts, 1788-89.
30. Duties on Beer, Malt and Hops.
31. Leases of the Blyth Ropewalks, 1762-1813.
32. Brickmaking accounts, 1774.
33. Blyth-built ships registered in Newcastle and Whitby, 1786-99.
34. Numbers of ships built in Blyth, 1750-99.

17

1. *Lease of the rights in the river Blyth, 1752.*

THIS INDENTURE made the first day of January in the Twenty fifth year of the Reign of our Sovereign Lord George the Second by the Grace of God of Great Britain France and Ireland King Defender of the Faith and so forth and in the year of our Lord one thousand seven hundred and fifty two BETWEEN the Right Reverend Father in God Lord Bishop of Durham of the one part and Matthew Ridley of the Town and County of Newcastle upon Tyne Esquire of the other part WITNESSETH that the said Reverend Father for and in Consideration of the Surrender of a Lease made by Edward late Lord Bishop of Durham to Richard Ridley of the Town and County of Newcastle upon Tyne Esquire bearing date the twenty ninth day of August one thousand seven hundred and thirty two and in Consideration of the Rents and Covenants herein after mentioned and for divers other Causes and Considerations him thereunto moving HATH devised granted and to Farm letten and by these presents for him and his Successors Doth devise grant and to Farm lett unto the said Matthew Ridley his Executors Administrators and Assigns ALL that his Anchorage Beaconage Wharfage and Plankage in the River of Blyth in the County Palatine of Durham as also all the Shipwrack of the Sea happening upon the Rocks and Shore of Camboys and Blyth (one Moiety to be collected for the use of the said Reverend Father and his Successors always excepted) TO HAVE AND TO HOLD the said Anchorage Beaconage Wharfage and Plankage in the River Blyth aforesaid and the Shipwrack of the Sea aforesaid (Except before excepted) unto the said Matthew Ridley and Executors Administrators and Assigns from the making hereof for and during the Term of One and Twenty years thence next ensuing and fully to be compleat and ended YIELDING AND PAYING therefore yearly and every year during the said Term of One and Twenty years unto the same Reverend Father and his Successors or to his or their Receiver General or Assignee for the time being at the Exchequer at Durham the Rent or Sum of Ten Shillings of lawful Money of Great Britain at the Feasts of the Purification of the Blessed Virgin Mary Pentecost Lammas and Saint Martin the Bishop in Winter by even and equal portions without Deduction or Abatement for any manner of Taxes or Assesses either by Act of Parliament or otherwise howsoever AND if it shall happen the said yearly rent of Ten Shillings or any part thereof to be behind and unpaid by the space of Twenty days next after any of the said Feasts at which the same ought to be paid as aforesaid That then and from thenceforth the present Indenture of Lease to cease and determine and be utterly void anything herein contained to the contrary not withstanding AND the said Matthew Ridley doth covenant and grant by these presents to and

19

with the said Reverend Father and his Successors That the said Matthew Ridley his Executors Administrators and Assigns shall and will from time to time during the said Term carefully collect and get in all the Shipwrack before mentioned and happening within the limitts aforesaid and shall and will faithfully every year give a perfect account thereof to the said Reverend Father and his Successors or his or their Receiver General or Assignee for the time being (when thereunto requested) and render to them either of them the full Moiety of the said Shipwracks and Profits thereof happening as aforesaid deducting only one Moiety of the necessary charges in the procuring and gathering the said Shipwracks AND the said Reverend Father for himself and his Successors doth covenant and grant to the said Matthew Ridley his Executors Administrators and Assigns that the said Matthew Ridley his Executors Administrators and Assigns shall and may quietly hold occupy and enjoy all and singular the demised premises with the appurtenances (Except before excepted) during all the Term aforesaid without any lawful disturbance or molestation from him the said Reverend Father or his Assigns or any other Person by his procurement IN WITNESS whereof the said Parties to these present Indentures interchangeably have set their Hands and Seals the Day and Year first above written.

(Signed by) Jo Duresme   (Joseph Butler, bishop of Durham, 1750-52)

Episcopal seal attached

(Witnessed by) Edwd Pearson N.P. (notary public), Willm Emm

2. *Two rent rolls of Newsham Estate, 1698.*

(1) A Rent Role of the Lordship of Newsham in the County of Northumberland scituated about 7 miles North of Newcastle.
1698

		£	s	d
West Farme part of it	To Mowe	60	0	0
The remainder	To Thomas Urwin	25	0	0
A slip of Groud betwixt that and the Demaine unlet ..		8	0	0
East Farme	To Young Midford & Jubb ..	90	0	0
Blyth Nooke Farme	To Laurance & Thornton ..	45	0	0
Phillip Jubbs house £2	Fishing £5	7	0	0
The Demaine		40	0	0

The Warren and 4 Closes	To John Blackett	..	..	30	0	0

Croswells Thistle Close ½ the Meadow Close & part of the high-
night Close 28 0 0

Croswells 2 great Closes & ½ the Meadow Close
                       To Jos: Browne..   ..   .. 30 0 0

New house att Blyth Nooke To Wm. Thornton 30 0 0

                                                        £393 0 0

(2) A halfe yeares Rent Roll of Newsham Estate due at Mayday 1698.

                                  £   s   d

South West Farme ⎰ Thomas Urwen / Robert Hunter / James Mitford ⎱ at 12 0 0 lease about 5 years

The great West Farme ⎰ Was let by Mr. Wyersdell with Covenant to build 2 farme houses 2 byers and 2 bar-nes but never built nor tenants never came so still in the Companys Hands, the Halfe years Rent was to bee ⎱ 30 0 0 No lease

Easte Farme ⎰ Richard Young / Edward Mitfor / Phillip Jubb ⎱ at 45 0 0 No lease

Blyths Nooke Farme ⎰ Thomas Loraine / William Thornton / John Blaket ⎱ at 22 10 0 lease about 5 years (erased)

The Warren and fower Closes ⎰ John Blaket ⎱ at 15 0 0 No lease

Cuthberts and Hymers Farme ⎰ Joseph Cuthberts and Cuth. Hymers ⎱ at 14 0 0 lease about 5 years

Creswells Closes ⎰ Now lett to Richard Young these two years ⎱ at 15 0 0 No lease

The Demeasne and slip of ground	formerly let to Mr. Wyersdell but Now to Mr. Erington at	at 21	0	0	Lease made for 9 years but not signed.
Great House at Blyths Nooke	To William Thornton for his .. saltery. Not let but valued to be worth £38 per Annum.	15	0	0	Not lett
Phillip Jubs House and ..	Phillip Jubb	1	0	0	No lease
Fishing	I suppose about £7 per Annum but I never received the Rent.	3	10	0	No lease

	£195	0	0	(for half year)
	£390	0	0	(for whole year)

3. *Survey of the Lordship of Newsham, 1785.*

Survey of the Lordship of Newsham in the Parish of Earsdon and County of Northumberland belonging to Sir Matthew White Ridley Bart.

Map No.	No.	Name of Fields	Quantity in each field A. R. P.			Quantity in each farm A. R. P.		
		Waggon Way Farm Timy. & Josh Dukesfield rent £130						
188	1	West Waggon Way Field	29	0	2			
189	2	Housesteads	23	3	32			
187	3	House Stackyards & Waste	2	0	21			
190	4	East Waggon Way Field	29	0	18			
215	5	East Middle field	24	1	16			
216	6	Laverick Hall field	14	2	8			
217	7	West Middle field (including Lane 0. 1. 22.)	14	3	9			
219	8	Stickley West field	28	3	8			
220	9	Newsham West field	27	3	19	194	2	13

Map No. No.		Name of Fields	Quantity in each field A. R. P.			Quantity in each Farm A. R. P.		
		Newsham West Farm	Timy. Dukesfield Junr. *rent £102 10s.*					
248	10	Houses Stackyards etc.	1	2	25			
247	11	Little Field	5	3	32			
249	12	Well Close	15	0	8			
250	13	Camp Field	19	3	30			
251	14	West Blakeburn..	12	2	5			
252	15	Borehole Field ..	13	0	33			
254	16	Middle Blakeburn	10	2	24			
253	17	Bauseys Field ..	11	3	10			
255	18	Great Field	18	3	3			
256	19	East Blakeburn Close ..	10	2	12	120	0	22
		Newsham Middle Farm	Joseph Clark *rent £100*					
258	20	West Blakeburn..	16	3	4			
260	21	East Blakeburn (including Bog 2. 0, 22.)	15	3	4			
259	22	Middle Field	18	2	20			
257	23	Shoulder of Mutton	16	1	8			
246	24	West Field	17	2	4			
245	25	East Field	11	1	14			
222	26	House, Yards, Garden and Waste	3	2	8			
221	27	West Greens	14	2	15	114	1	37
		Newsham East Farm	William Bennett *rent £100*					
	28	House, Stackyard, Lane and Waste ..	3	3	2			
224	29	Bog Field	10	3	13			
225	30	Shoulder of Mutton	14	1	30			
226	31	Bog Field	12	1	2			
244	32	Broad Field	14	3	20			
263	33	Horse Pasture ..	10	1	26			
264	34	Long Close	17	0	6			
262	35	West Blakeburn..	19	0	27			
266	36	Middle do. ..	16	2	24			
268	37	East do. ..	10	3	21	130	1	11

Map No. No.		Name of Fields	Quantity in each field A. R. P.			Quantity in each Farm A. R. P.		
		Waggon Way East Farm John Watson *rent £120*						
211	38	Horse Pasture	15	1	2			
212	39	Broad Flatt	16	1	15			
213	40	Newsham Field	22	3	26			
214	41	Rose Pasture	11	2	30			
191	42	The March	7	1	8			
186	43	Stackyard Garden and Waste ..	1	3	7			
192	44	Garden Field	16	3	14			
193	45	Broad do.	19	2	15			
194	46	Bean do.	11	3	24			
210	47	Middle do.	9	1	35			
228	48	Link do.	9	3	26	143	0	12
		Link Farm, Ellstob & Hog *rent £120*						
227	49	Near Barley Close	22	3	2			
235	50	Far Barley do.	14	1	20			
233	51	Priors Close	9	1	21			
232	52	South Link Close	15	3	1			
230	53	North do.	15	2	0			
229	54	High Close	20	1	8			
209	55	Shottons do.	11	2	23			
208	56	Shed Field	18	1	28			
			128	0	23			
205	57	Link Pasture	33	1	6			
207	58	Clay Holes	5	0	0	166	1	29
		Blyth Mrs. Marshall						
198	59	Low Close	6	3	30			
197	60	Low Middle Close	11	0	34			
196	61	High do.	7	0	32			
195	62	High Close	8	0	4	33	1	20

Map No. No.		Name of Fields				Quantity in each field A. R. P.			Quantity in each Farm A. R. P.		
		Blyth Messrs. Clarke & Watt *rent £34*									
199	64	South West Field	..	..	..	4	0	24			
203	65	South East do.	..	..	..	3	0	24			
202	66	North East do.	..	..	..	3	3	4			
201	67	North West do.	..	..	..	3	0	20	13	3	32
184	63	Mr. Jackson's Close (*rent £10*)			..				7	0	2
		Wastes									
206 &		The Links	..	..	..	..	54	2	17		
231		Ballast Hill	..	..	..	..	5	2	26		
		Waggon Way	..	..	..	..	10	0	20		
		Blyth Town	..	..	..	..	16	0	4	86	1	27
		Grounds in hand									
200	68	Waggon Way Field	..	..	..	6	3	27			
204	69	Ropery do	..	..	..	7	2	15	14	2	2
		Link House Farm, belonging to Nicholas Ridley Esqr. Margt. Dobson, Tenant									
237	70	Little Close	..	..	..	6	1	26			
239	71	Back of Garden..	..	..	..	1	0	32			
234	72	Broad Pasture	..	..	..	23	3	20			
236	73	Burn Close	..	..	..	9	0	20			
241	74	Link House Close	..	..	..	13	2	26			
243	75	Newsham Field	..	..	..	10	3	20			
265	76	West Clover do.	..	..	..	12	3	22			
269	77	East Clover do.	..	..	..	13	3	7			
271	78	Link Close	..	..	..	11	0	18			
242	79	Planting in do. ..	..	..	..	0	0	25			
270	80	South Intake	..	..	..	5	2	14			
267	81	Burn Side	..	..	..	5	1	26			
272	82	South Links	..	..	..	23	0	31			
238	83	North do.	..	..	..	16	2	24	154	0	31

Total of the Lordship   1,178   1   38

4. *Tillage of Newsham Estate, 1787*

Land in Tillage in Newsham Estate 1787

	Farms	Grain Sown	Quantity			Total in each Farm.		
			A.	R.	P.	A.	R.	P.
	Watsons Farm							
1.	Broad flat	Wheat	16	1	0			
2.	Rose Pasture	Wheat	3	0	0			
	do.	Barley	2	0	0			
	do.	Beans	7	0	0			
3.	Garden Field	Oats	16	3	14	45	0	14
	Tim. & Jos. Dukesfield							
		Fallow {	14	1	27			
			13	0	0			
1.	West Field	Oats	14	0	0			
2.	Laverick Hall field	Wheat	14	2	8			
3.	Stickley West field	Oats	28	3	8	57	1	16
	Timy. Dukesfield Newsham							
		Fallow {	10	0	0			
			12	2	5			
1.	Newsham West field	Wheat	23	3	19			
	do. do.	Barley	4	0	0			
2.	Borehole field	Wheat	5	0	0			
	do. do.	Barley	8	0	0			
3.	Well Close	Oats	15	0	8			
4.	Bassey's Close	Oats	11	3	10	67	2	37
	Clark & Wilson							
	East Blakefield Westfield	Fallow {	13	0	0			
			9	0	0			
1.	Middlefield	Oats	15	0	0			
	do.	Beans	3	0	0			
2.	Shoulder Mutton field	Wheat	16	1	8			
3.	West field	Barley	2	0	0			
	do.	Oats	6	0	0	42	1	8

26

Farms	Grain Sown	Quantity			Total in each Farm.		
		A.	R.	P.	A.	R.	P.
Willian Bennett							
	Fallow	24	0	0			
1. Shoulder Mutton field	Wheat	2	0	0			
do. do.	Oats	4	0	0			
2. West Blackburn	Wheat	19	0	27			
3. Middle Blackburn	Oats	16	2	24	41	3	11
Margt. Dobson							
Broad Pasture	Fallow {	12	0	0			
East Clover Field		8	0	0			
1. Broad Pasture	Barley	11	0	0			
2. Newsham Field	Beans	7	0	0			
do. do.	Oats	3	0	0			
3. West Clover field	Oats	5	0	0			
4. East do.	Oats	7	0	0			
5. Linkhouse Close	Wheat	10	0	0	43	0	0
Hogg & Elstob							
Far Barley field	Fallow {	14	0	0			
Highfield		8	0	0			
1. South Link Close	Wheat	15	3	1			
2. North do.	Wheat	15	2	0			
3. Highfield	Oats	12	0	0			
4. Shade field	Beans	5	0	0			
do. do.	Oats	13	0	0	61	1	1
				Acres	358	2	7

Total quantity of each sort of grain in
Newsham Estate 1787.

		A.	R.	P.
	Wheat	141	1	23
	Barley	27	0	0
	Beans	22	0	0
	Oats	168	0	24
	Acres	358	2	7

27

5. Cropping of Link House Farm, 1812-15

No.	A.	R.	P.		1812	1813	1814	1815
73	9	0	20	⎱ 8	Fallow	Wheat	Clover	Wheat
74	13	3	26	⎰	Clover	Oats	Fallow	
75	10	3	20	⎱	Wheat	Clover	Beans	Wheat
76	12	3	22	⎰ 5¾	Clover	Oats	Fallow	
77	13	3	7		Wheat	Clover	Oats	Clover
78	11	0	18	⎱ 6	Oats	Fallow	Wheat	do.
				⎰	Oats	Fallow	Oats	Oats
79	5	2	14	⎰ 5	Fallow & Turnips	Oats	Fallow & Turnips	
					Clover		Fallow/ Seeds	
	77	1	7					

6. *Profits from the Blyth and Newsham Estates and Trade, 1789-99.*

Year or period ending	Colliery £	s.	d.	Newsham Estate £	s.	d.	S. Blyth Houses £	s.	d.	Salt Pans £	s.	d.	Brewery £	s.	d.	Other £	s.	d.	Total £	s.	d.
31.12.1788-22.11.1789	3466	10	9¾	827	0	10¾	335	13	3½	2	18	0¼	274	9	10				4906	12	9¼
22.11.1790	4385	12	2	850	9	2	450	10	8¼	58	17	8¼	284	14	5				6030	4	1¾
22.11.1791	4612	7	2	790	0	6	392	7	8¾	121	15	10¾	301	14	9				6218	6	0½
22.11.1792	3870	3	1¼	809	5	4	148	4	6	37	10	8½	262	13	6½				5203	7	1¾
																75	9	11[1]			
22.11.1792-12. 5.1793	1434	12	11	468	11	5½	201	7	9	45	17	8½	238	9	0½				2388	18	10½
12. 5.1794	2921	5	6¼	880	4	9½	440	3	6	225	8	9½	350	17	6½	790	0	0[2]	5608	0	1¾
12. 5.1795	3539	0	3½	845	2	11	125	7	2	181	4	11¼	379	6	10				5070	2	1¾
15. 5.1796	3308	6	2¼	872	19	11½	26	15	10½	206	3	1½	564	0	3	400	0	0[3]	5378	5	4¾
12. 5.1797	389	1	7¼	927	8	0	391	8	1½	202	11	6	677	10	0				2587	19	2¾
12. 5.1798	2232	11	0½	854	12	4½	444	16	11½	196	8	6½	423	8	10¼				4151	17	9¼
12. 5.1799	987	2	4	888	6	4	525	7	2½	217	15	10½	497	3	5				3160	15	2
12. 5.1799-31.12.1799	528	9	10	384	15	5	207	18	11½	156	11	0	271	3	7				1602	18	9½

1 Timber from River Green

2 3 Cash from Sir M. W. Ridley

7. Coals wrought from Bassington Colliery pits, 1783-88.

Year	West Pit		South Pit		Rodney Pit		Rising Sun Pit		Hall Pit		Success Pit		Total		Value		
	Tens[1]	Wgns.	Tens	Wgns	Tens	Wgns	Tens	Wgns	Tens	Wgns	Tens	Wgns	Tens	Wgns	£	s.	d.
1783	357	0	396	11	423	11	—	—	—	—	—	—	1,177	0	4453	14	0 [2]
1784	—	—	127	0	510	0	532	11	—	—	—	—	1,169	11	4385	18	1½ [2]
1785	—	—	355	0	477	0	453	0	32	0	—	—	1,317	0	4874	12	0 [2]
1786	—	—	398	0	458	0	—	—	194	0	—	—	1,050	0	3092	16	3 [2]
1787	—	—	45	11	599	11	334	11	42	3	152	11	1,174	0	4738	5	5½
1788	—	—	—	—	348	0	168	11	45	11	591	0	1,117	0	4057	14	4

[1] The Ten was a measure of coal used in Northumberland and Durham upon which a colliery lessor's rent was fixed. The actual amount varied in the Blyth area from 48·33 to 48·583 tons. The entries make it clear that 22 waggons were equal to one ten.

[2] These years include payments for extra price of oats: £50 in 1783 and 1784; £84. 0. 6. in 1786.

8. Coals led to Blyth Staith and their disposal, 1788.

Dr

Leading Coals to Blyth Staith:

	Tens	Waggons
West Pit	180	21
Rodney Pit	301	7
Success Pit	503	20
Total	986	4
(Value) £1356.0.0		

Cr

And How the Coals are Disposed of:

	(Waggons)
Round & Mixed Coals to Ships	17228
Brewery Coals delivered	39
Small Coals to Ships	2185
Ching Coals to Town	123
Ching Coals Struck	292
Small Coals: Town and Country	254
Small Coals to W. Carr	25
Small Coals to Trimmers	125
Ching Coals to J. Turner	43
Small Coals to Jos. Wheatley	30
Small Coals to Brewery Men	11
Small Coals to Trimmers Widows	19
Small Coals to No. B. Pans	10
Small Coals to So. B. Pans	1293
Small Coals to J. Clark	15
Small Coals to Lime Kiln	6
	21698

Total 986 Tens 4 Waggons

Note:—The figures of the disposal have no indication of the unit of weight but the total of 21698 appears to be of waggons since a figure of 22 waggons to the ten would give a total of 986 tens 6 waggons. If we accept the figures of coals wrought at various pits (7) it appears that Rodney produced 348 tens and led 301, Success

31

produced 591 and led 503 but West Pit which was not shown as producing any coal in 1788 is shown as leading 180 tens. Could it be that the production of 168 tens from Rising Sun and 45 tens from Hall were both recorded at Blyth as coming from West Pit which the other record shows as ceasing production in 1783?)

9. Cost of maintenance of the Blyth, or Plessey, Waggon Way, 1784.

Blyth Waggon Way

1784	To Cash paid to:	£	s	d
Jan 23	Wrights and Labourers Account this fortnight per Journal	12	14	2
Feb 6	Wrights and Labourers Account this fortnight do. Journal	6	0	0
20	Wrights and Labourers Account this fortnight per Journal	5	15	10
Mar 5	Wrights and Labourers Account this fortnight per Journal	5	19	2
19	Wrights and Labourers Account this fortnight per Journal	5	19	7
	To the Rt. Hon. Lord Delaval for Half years Wayleave Rent	150	0	0
24	Wm. Row for Waggon Rails freight per Journal ..	44	11	0
April 2	Wrights and Labourers Account this Fortnight do.	5	19	4
April 16	Wrights and Labourers Account this Fortnight per Journal	5	18	8
30	Wrights and Labourers Account do. do. ..	5	18	8
May 10	Robert Harrison for 12 Metal Wheels per Journal	19	14	11
14	Wrights and Labourers Account this fortnight per Journal	6	0	0
19	John Watts for Ballasting the Way and Labour Allowance	6	5	2
	William Carr for Halfyears Tentale at the Waggons do.	15	14	9
	John Watts for Halfyears Tentale at do. do. ..	16	14	9
	Robert Dobson for Ballasting the Way Halfyear ..	2	10	0
	Matthew Gray for Ballasting the Way do. do.	5	0	0

Date		Description	£	s	d
May	28	Wrights and Labourers Account this Fortnight do.	5	17	6
June	11	Wrights and Labourers Account this Fortnight do.	5	16	0
	25	Wrights and Labourers Account this Fortnight do.	6	0	0
July	2	Freight of Oak plank and Sleepers from Cresswell Sands	3	3	0
	5	Jane Marshall for Deals etc. as per Journal ..	15	16	6
	9	Wrights and Labourers Account this fortnight per Journal	6	0	0
	23	Wrights and Labourers Account this fortnight per Journal	6	0	0
Aug	6	Wrights and Labourers Account this fortnight per Journal	6	0	0
	16	freight of a Sloop Load of Rails etc. do. ..	2	2	0
	20	Wrights and Labourers Account this fortnight do.	5	19	2
Sepr	3	Wrights and Labourers Account this fortnight do.	7	0	0
	13	Rt. Hon. Lord Delaval for Halfyears Way Leave Rent	150	0	0
	17	Wrights and Labourers Account this fortnight per Journal	5	17	4
Oct.	1	Wrights and Labourers Account this fortnight per Journal .. .. ,,	6	0	0
	15	Wrights and Labourers Account this fortnight per Journal	5	19	8
	29	Wrights and Labourers Account do. do. ..	5	18	4
Nov	12	Wrights and Labourers Account this fortnight per Journal	6	0	0
	18	William Row for Sloop Load of Beach Rails and Wheels per Journal	19	19	9
	26	Wrights and Labourers Account this fortnight per Journal	6	15	9
	30	Robert Dobson for Ballasting the Way Halfyear do.	2	10	0
	30	William Carr for Halfyears Tentale of the Waggons do.	22	15	4
		John Watts for Halfyears Tentale of the Waggons do.	21	15	4
		Matthew Gray for Ballasting the Way Half a Year do.	5		
Dec	2	John Watts for Ballasting the Way Half a Year do.	8	4	7
	10	Wrights and Labourers Account this fortnight as per Journal	10	7	6
	16	William Atkinson for Waggon Rails vended to Plessey Colliery do.	122	0	10

33

Trewick and Pearson for Sleepers vended to Plesesy
Colliery as per Journal 47 19 11
23   Wrights and Labourers Account this fortnight as per
Journal 5 18 6

10. *Payments of maintenance for the Waggon Way, 1774.*

1774

May 30 There have been led from Hartford Pits from Xmas
to & with the 12th instant ... 327$\frac{3}{22}$ Tens of coals £   s   d
Due thereon to Waggonmen for Leading
£1 7. 6.   ..   ..449 12 6

Robert Robson	at 2s	..	..	..	.. 32 14 3
Jno. Watts	9d	..	..	..	.. 12 5 4
Wm. Carr	11d.	..	..	..	.. 14 19 10½

Dr.	£	s	d		Cr.	£	s	d
Staith	32	14	3		R. Robson	32	14	3
Wag. Way	12	5	4		Jno. Watts	12	5	4
Wag. Way	14	19	10½		Wm. Carr	14	19	10½

June 6 Due on the 12th May to Wm. Carr for repairing the
Waggons for ballasting the Way 2 0 0
Three new Axle trees 0 15 0
Lock & Key mending for Schoolhouse 0 1 0
_____
2 16 0

_____
£   s   d
June 24 Mr. Crooks Bought of Jno. Grey on the 14th instant, viz

346 yds of Ash Rail		5d.	..	..	.. 7 4 2
373 Oak sleepers		7d.	..	..	.. 10 17 7
6 soles	at	2s.	..	..	.. 12 0
60 Ash Sleepers	at	5d.	..	..	.. 1 5 0
8 Doz. side sheaths		4s.	..	..	.. 1 16 0

_____
21 14 9

June 27 R. S. has paid Robert Reed in full for leadings to the
    12th May 10 14 6
    And has kept off for breaking a metal wheel 31st of
    Jany. 0 10 6

                                           11 5 0

				£	s	d

Nov 25 Since Midsummer Jno. Grey has delivered at Plessey

			£	s	d
429 yds Oak Rails	at 6d.	10	14	6
615 Do. Ash Rails	at 5d	12	16	3
10 Doz. side sheaths	at 4s. 6d.	2	5	0
6 Bottom sheaths	2s.	0	12	0
47 Oak sleepers	at 7d.	1	7	5
120 Ash sleepers	at 5d.	2	10	0
30 Soles	2s.	3	0	0
44 Overings	1s.	2	4	0

                                           35 9 2

    To which R. S. hath paid him by draught on Surtees
    & Burdon 35 9 2

11. *Cost of maintenance of Blyth Staith, with expenses for office and keels, 1784.*

    Blyth Staith etc.

1784	To Cash paid to		£	s	d
Jan 5	Geo. Tate for painting			8	10½
8	John Elliot	Carpenter work ..	10	8	6
10	Agents wages		5	19	0
,,	Charnley	Paper & Books ..	4	9	10
,,	Heatherington	Stamps etc. ..	5	2	6
17	John Elliot	Carpenter ..	4	0	0
24	Agents wages		5	4	0
Feb. 7	do.		5	4	0
19	Cleaning office ¼ year			2	6
21	Agents wages		5	4	0

			£	s	d
Mar 4 Geo Brown	Smith work	.. ..	10	19	2
6 Agents wages	.. ..	.. .. ..	5	4	0
13 John Elliot	Carpenter	.. ..	4	3	0
„ Wm. Weatherhead	Mason..	.. ..	4	13	8
20 Agents wages	.. ..	.. .. ..	5	4	0
„ Filling Rubbish	..	.. ..		8	4
31 Bart. Kent	Paper etc. for Office			14	9
„ Leading Oak Plank	..	.. ..	1	7	9
„ Postage of letters, etc. ..		.. .. ..	4	3	1
Apl. 3 Agents wages	.. ..	.. .. ..	5	4	0
15 Postage of letters etc. ..		.. .. ..		6	7
17 Agents Wages	.. ..	.. .. ..	5	4	0
30 Postage of letters	..	.. .. ..		11	2
May 5 Joseph Gray	Keel sails	..	12	11	6
15 Agents Wages	.. ..	.. .. ..	5	4	0
„ Stamps for Receipts	..	.. .. ..	1	10	0
19 John Watts	Allowance etc.	..	6	11	0
25 John Elliot	Carpenter	..		19	7
„ John Johnson & Co.	Oak Plank etc.	..	24	11	10
29 Agents wages	.. ..	.. .. ..	5	4	0
„ John Elliot	½ years Account	..	8	15	10
„ Geo. Brown	Smithwork	.. ..	8	13	0½
„ Robt. Dobson	Conveying Ballast	..	41	19	4
June 12 Agents Wages	..	.. .. ..	5	4	0
„ Carpenter & Smiths Work	.. ..	.. ..	1	16	10
26 John Elliot	.. ..	.. .. ..		4	0
„ Geo. Brown	.. ..	.. .. ..		7	4
„ Agents Wages	.. ..	.. .. ..	5	4	0
July 2 Freight of Oak Plank ..		.. .. ..	3	3	0
5 Jane Marshall	Timber	..	16	17	5½
10 Agents wages	.. ..	.. .. ..	5	4	0
„ Geo. Brown	Smithwork	..		2	8
„ John Elliot	.. ..	.. .. ..		3	2
„ John Marshall	Ropes	..	13	13	10½
20 Bielby & Co.	Office seal	..		12	0
24 Agents wages	.. ..	.. .. ..	5	4	0
„ John Elliot	Carpenter	..		7	0
„ Geo. Brown	Smith	..	1	19	11
Aug. 7 Agents wages	.. ..	.. .. ..	5	4	0
„ Geo. Brown	.. ..	.. .. ..		2	8
14 Saml. Aislabie	¼ Years wages	..	16	5	0

				£	s	d
Aug	21	Agents wages		5	4	0
„		John Elliot		1	4	6
„		Geo. Brown			19	1
	23	John Elliot		1	19	10
„		Geo. Brown		8	13	0
Sept	2	J. Scaife's expenses Sunderland			8	4
	4	Agents wages		5	4	0
„		John Elliot	Carpenter		13	8
„		Geo. Brown	Smith	1	2	5
„		J. Scaife's expenses at N'castle			2	6
	18	Agents wages		5	4	0
„		Geo. Brown			7	1½
	25	2 Keels Setts			3	6
	28	Postage of letters		4	14	0
Oct	2	Agents wages		5	4	0
„		John Elliot	Carpenter		10	0
„		Geo. Brown	Smith	1	3	7
	16	Agents wages		5	4	0
„		Casting Rubbish		1	1	10½
„		Geo. Brown	Smith	1	0	7½
„		John Elliot		2	5	0
„		Wm. Weatherhead	Mason	6	5	10
„		J. Scaife's expenses at Sunderland			3	6
	21	3 Keels Setts			6	0
	30	Agents wages		5	4	0
„		John Elliot		1	10	6
„		Geo. Brown		2	14	7
Nov	12	4 Keel Setts			8	0
	13	Agents wages		5	4	0
„		Geo. Brown		1	15	11
„		Jacob Bell	Oak plank etc.	2	3	3
„		Agents wages		5	4	0
	27	John Elliot			4	9
„		Geo. Brown			4	1½
Dec.	11	Agents wages		5	4	0
„		Mark Watson		40	12	6
	20	Hull of the *English Nero*		22	0	0
„		Robt. Dobson	Conveying Ballast	58	0	10
„		John Watts	Allowance	4	16	4
	24	Agents wages		5	4	0
„		Leading Oak Plank		2	18	0

							£	s	d
„ do. do.	..	..	..	..	..	..	1	0	0
„ do. do.	..	..	..	..	..	..		15	0
„ do. do.	..	..	..	..	..	..		10	0
„ John Elliot	Carpenter				..	..		13	0
29 taking up an Anchor	..		..	..	..	..	1	1	0
31 Wm. Charnley	Books etc.			..		..	4	1	9
„ Leading Wreck etc.	..		..	..	..	..	10	7	4
„ Jane Marshall	..		..	..	..	..	24	6	0
„ John Marshall	Ropes ..			..		..	2	15	5

£560 4 11

Note. This account is continued on four separate pages in the ledger. The first page is headed as above, the second page is headed "Blyth Staith, Office, &c", while the last two are headed "Blyth Staith, Keels &c."

12. Description of Blyth Harbour by Sir John Rennie 1814.

The Town of Blyth is situated between 2 and 4 miles north of Seaton Sluice. Its harbour is formed by a small river of the same name which at its entrance into the sea is in some degree sheltered on the north-east by a long ledge of rocks called Coble Hole Rocks, which are about three-quarters of a mile in length, and which may be said to form the northern extremity of Seaton Bay. These rocks are covered at high water but are dry at low water. To the seaward lyes a cluster of rocks called Seaton Sears and north-east of these is another cluster of rocks called the Sow and Pigs.

These rocks are dangerous to vessels sailing out of or coming into the harbour in southerly winds and the shallowness of the water and exposed situation of its entrance render it quite inaccessible in stormy weather, particularly in northern and north-easterly winds. The Blyth river is very small and the bottom is hard from a little west of the Basket Rock to near its mouth at low water, so that it is unable to scour and maintain a deep channel; besides the sand on the south flat and soft and the seas from the north break across Coble Hole Rocks with great violence, and throw such a quantity of stones into the channel that they obstruct the current and occasion it to spread to the southward, so that there is not above 10 inches or a foot depth at low water while abreast of the quay there is upwards of six feet. No vessels can therefore get into the harbour in Neap tides unless they are very small and but moderate sized vessels can get into it on Spring Tides when wind and weather serves; but as otherwise they are frequently detained until the Springs are gone and they have to wait for another Spring.

13. *Monthly numbers of ships loaded at Blyth, 1755-67 and 1793-99.*

Year	Jan	Feb	Mar	Apl	May	June	July	Aug	Spt	Oct	Nov	Dec	Total
1755	7	3	22	18	20	25	18	23	17	26	10	11	200
1756	14	10	7	13	25	31	16	16	19	17	16	16	200
1757	8	15	14	20	22	23	19	21	11	10	10	3	176
1758	8	6	8	15	11	13	20	13	9	10	14	9	136
1759	16	7	5	12	14	18	11	10	13	11	12	10	139
1760	7	2	9	15	15	16	14	22	15	10	6	7	138
1761	12	8	16	18	22	19	23	21	14	9	7	8	177
1762	8	4	6	11	22	16	21	13	14	9	13	13	150
1763	6	8	14	9	15	25	17	18	10	19	8	7	156
1764	10	10	17	14	24	19	22	21	14	19	7	8	185
1765	9	9	8	11	25	22	19	13	10	11	10	11	158
1766	3	6	9	24	21	24	32	15	12	15	7	3	171
1767	1	9	3	12	19	17	25						
Average	8·3	7·4	10·6	14·4	19·4	20·6	19·7	17·1	13·1	13·8	10·0	8·3	165
1793	26	25	27	27	28	35	23	28	25	29	23	32	343
1794	23	24	31	36	28	23	35	28	20	27	24	16	315
1795	6	3	22	25	31	40	50	24	30	32	18	17	298
1796	18	15	12	32	32	35	38	30	19	26	9	2	267
1797	16	20	22	15	38	31	31	26	21	19	16	22	277
1798	19	12	15	24	24	23	27	26	25	22	19	6	242
1799	22	3	7	11	28	24							
	18·5	14·5	19·4	24·2	34·1	30·1	34·0	27·0	23·3	25·6	18·1	15·8	292

14. *Registration of ships taking coal cargoes, 1765 and 1797.*

1765 Port to which ship belongs	Number of cargoes	1797 Port to which ship belongs	Number of Cargoes
Lynn	70	Lynn	55
Blyth	31	Blyth	123
Scarborough	16	Scarborough	10
Whitby	11	Whitby	18
Shields	10	Shields	10
Hartlepool	3	—	—
Newcastle	2	—	—
Sunderland	2	Sunderland	22
Wisbeach	2	Wisbeach	3
Wells	2	Wells	5
Yarmouth	2	Yarmouth	5
Ramsgate	1	Seaton	5
Southwold	1	Hartley	5
Milford	1	Rochester	3
Ipswich	1	Woodbridge	2
London	1	Lyme Regis	2
Newbiggin	1	Cromer	2
		Boston	2
		Perth	1
		Ramsgate	1
		Bridlington	1
		Hastings	1
		Stockton	1

15. *Coal cargoes carried by Blyth ships, 1754-62 and 1790-98.*

Year	Ships	Cargoes	Year	Ships	Cargoes
1754	2	2	1790	24	87
1755	2	4	1791	29	164
1756	3	6	1792	26	147
1757	11	27	1793	28	144
1758	7	18	1794	25	123
1759	8	22	1795	23	133
1760	9	27	1796	23	104
1761	9	28	1797	26	123
1762	6	19	1798	28	132

16. *Coals shipped and harbour dues at Blyth, 1783-88.*

Year	Ballast, Conveying & Harbour Dues			Best (or) Round Coals			Mixed Coals			Small Coals		Total		
	£	s.	d.	16/6	16/-	15/-	14/-	12/-	11/-	6/-	5/-	£	s.	d.
1783	125	1	0	—	5627	424	—	—	5011	822	—	8261	5	6
1784	139	17	6	—	6442	228	—	—	6463	1147	—	9221	19	0
1785	150	16	0	—	7330	184	—	7528	—	1099	—	10836	8	6
1786	125	19	0	—	6802	274	—	5637	—	1507	—	9479	5	0
1787	163	9	0	—	5931	361	—	5818	—	1607	—	10433	19	11
1788	154	18	0	400	4762	42	98	8183	—	1780	93	9756	18	1

Note: The coals are measured in chaldrons of 53 cwt.

D

17. *Letters from Matthew Ridley relating to the Coal Trade, 1767-75.*

Geo. Ward Esq.   Newcastle, Nov 6 1767

I am favoured with your letter of the 2ᵈ and was in hopes that as winter now comes on and many people of Business hurrying to town, that the Demand for Coals wou'd have been more brisk, and certainly it must be so unless the Stocks on hand be extraordinary large, I shou'd hope that the Blyth Coals wou'd be called for, you will observe in engaging a Ship for those Coals not to be of a Burden above 6 or 7 Keels, if I meet with one convenient here shall despatch it, On the other side you have the Ships laden with Byker and T.M.E. since my last. The weather is mild and the wind S.W. so the Ships will not make a quick passage.

M.R.

Geo. Ward Esq.   Newcastle, Jan 22 1768

I wrote to you on the 16th. Since which I am favoured with yours of the 18th, and am sorry that Darkin and the other two Ships you mention remain unsold, if they are not ere this Sold I am very doubtfull they will meet with a bad Market as it is imagined here that the large Fleet which sailed from Shields last Sunday will be now arrived in London River and another Fleet that sailed from thence last Tuesday will now be in yarmouth Roads where its thought they may be detailed awhile as the wind is now full South and blows fresh. We have here Clod, Shabby, dirty weather, if the same with you, it will increase the consumption of Coals, and make the Demand more Brisk, there are Scarce one Ship in ten that have Load since Xmas but what are upon Freight which is indeed a very disagreeable way of Vending Coals; if any other method cou'd be found out, but so long as Coals remain so very Plenty here I am doubtfull that it not be Totally laid aside. The freight at 9sh. a chaldron continues, which I am afraid are more than the price of Coals at Market will pay . . . .

D. Lamb

Geo. Ward Esq.   Newcastle, June 10 1768

I am favoured with your letters of the 3ᵈ. and 16th, the attempts of the Sailors altho at present attended with opposition & Disturbance from the Coal heavers, will I think bring the latter to bear, that they will work at reasonable prices, and the Sailors will soon be weary of such toilsome work, so that the Trade may revert to its own chanel, yet all this will depend on this Magistrates exerting themselves to punish the wrong Doers either Sailors or Coal heavers; I think the Trinity house with their authority & Influence will soon knock on the head the opposition of the Coal-heavers to the Ballast men.

Little encouragement here in the Coal Trade, the Masters dont know what to do, they run to the preferable sorts only, and there appears no Sign of amendment as to the owners, it cannot last long in this shape, as every man must be sick of throwing away his property in such a foolish way. On the other side you an acco't of what has been done as to my Coals. I shall go to Blyth today, and tell the Masters there what you say, surely 27sh for those Coals would turn out more profit than loading at Newcastle, the Expenses at the Port being so much lower; yet I do not care to freight; the Coals are extremely good.

I have now a large Quantity of Salt there and shou'd be glad to deliver 2 or 300 tons in a month even at 26sh per ton; money being at this time a very necessary Commodity; I sho'd be glad if you cou'd help me forward in this matter.

M.R.

Mr. Edmd. Elsdon   Newcastle, Feby 15 1769

Lynn Rs

Yours of the 11th. I have received, we always have small coals at Blyth or can lead them immediately the price is seven shillings a Chaldron, and if you choose round along with them twill be best to put the small underneath as you can take the roundest out first to answer the purpose you want them for, the objection to mixing them is to prevent bringing an ill character upon the Blyth Coals, I presume the turn is the same small or round, I am etc.

J.G.

Mr. Chas. Wildbore
   at the Trinity House
      London   Newcastle, Aug 11 1769

I am favoured with your letter of the 5th with Copy of a Representation from Mr. Even Johnson as to a Cargo of Coals shipped at Blyth for Scilly Light, I must confess myself extremely surprised at the Complaint, as Blyth Coals are universally acknowledged to be of a bright Quality for burning, and remarkable for their Roundness, for the Truth of which I appeal to my Friend Sir John Major; I went yesterday to Blyth to examine into the particulars of this Cargo and was assured by the three Servants that attend Loading the Ships that this Cargo ordered by Mr. John Wilson in May last, was as large as they ever Saw, in so much that the Master of the Ship said he shou'd have Difficulty in delivering the Coals by the Shovel, my Servants are ready to make Oath of the same; I therefore am persuaded that some Tricks have been played after the Ship sailed from Blyth, and it behoves me to be anxious to have the matter most strictly enquired into, as my Character is attacked in putting an inferior Commodity on a Buyer at the best price, I shou'd be much obliged to you if you wou'd examine into this matter as far as possible, and let me know how it turns out, I am with due Regards to the Corporation of the Trinity House.

M.R.

Geo Ward Esq   Newcastle, Oct 12 1770

I see by your favour of the 8th the presint Fluctuation in the Coal Trade, if War is the point it will not in the main hurt the Trade, I don't See how War can be avoided, if any Credit can be given to Accounts in the Gazette Herewith you have what has been done at Byker, I can say nothing for Blyth as so little Encouragement is given by the London Buyers, so you see only a Ship or two will run the Risque, which is really Surprising as the price by your Account is 33 and the Cost at Blyth but 6/6^d the London chaldron, but Hartley being 6^d above discourages.

I. am
M.R.

Geo Ward Esq   Newcastle, November 10 1770

The violent Rains on Thursday prevented that Post from getting hither till yesterday afternoon, I fear we shall hear of Damage at Sea two Ships, one of Lynn, the other Whitby are put on Shore near Blyth, but as the weather was moderate yesterday, they hope to get them off without Damage, I perceive by your Favour of the 5th that the prices keep up pretty well, we have despatched with Byr. the two Ships on the other Side; notwithstanding what you mention of Blyth Coals selling well even Hannay's Ships go from Blyth to Hartley which is very discouraging even to Strangers, I wish to be informed the Reason for it, for the Coals are as good as can be, and the price the same as Hartley, and the measure better, I must talk to you about it when we meet, which will I hope be by this day Week, I was prevented setting forward so soon as I first intended, The ship for Mr. Bowles Salt cannot conveniently take in more than 150 tons of Salt, but shall be despatched with that Quantity on her arrival at Blyth.

I am Yours
M.R.

George Ward Esq   Newcastle, June 14 1771

I am favoured with your letter of the 10th, what rumoured as to the
Pitmen and Keelmen stoping took place on Tuesday last, the Pitmen are
assembled in great numbers and proceed from Colliery to Colliery to put
a stop to all Work, they are at present at the Colliery on the Sunderland
River, their outcry is the high price of Corn which I am afraid at this time
cannot be corrected; I shou'd hope that all persons concerned in the Trade
wou'd be of Opinion to relieve the necessities of the men so far as they
have it in their Power, if they behave peaceably and return to their Work.
The keelmen complain of loading Ships directly out of the Waggons by
Spouts without making use of Keels; this has been the practice from the
beginning of the Colliery at Chirton and Tinmouth Moor, I have just now
erected a Spout at Byker for small Coasting Ships as the Trade has un-
doubtedly a claim to be excused the Expense of Keel hire when they can be
better dispatched without, both in Saving the Coals from being so much
broken, and also dispatched, there is no saying, how long this Behaviour
of the Workmen may Continue, so no doubt the Ships at London will
take the Advantage. I expect that you will have a Ship or two at Blyth in
a few days . . . .

Geo Ward Esq   Newcastle, Aug 28 1772

I am sorry to See by your favour of the 24 that the freighted Ships from
Blyth have turned out to Loss the Freight Surely must have been very
high as the Coals were Sold at a good Price, but there was an Expense
which ought not to have been by the Master freighted taking out half their
Loadings to Sea in Keels which they wou'd not have done on their Own
Account, however the Affair is over and I must submit to the Loss, I
desire that you will acquaint me with the deficiency to reimburse you . . . .

Geo Ward Esq   Newcastle, Sept 3 1773

. . . . the Demand for the Coast in Small Ships which are
loaded from the Spout by the Waggons immediately Shooting into the
Ships makes the Coals better than when put into Keels, ...

                                                    M. Ridley

Geo Ward Esq   Newcastle, December 17 1773

I am glad to see by your Favour of the 13th that the Price of Coals have advanced and that may probably increase, as there has been very few Ships dispatched from this Port nor will be I take for granted untill after the Holidays; we have done nothing at Byker I do know what has been done at Blyth. I heartily wish you and your family many happy Returns of the approaching season and am

Yours etc.

M. Ridley

Geo Ward Esq   Newcastle, July 29 1774

I am favoured with your Letter of the 25 with the price of Coals; what has been done at Byker you have below, Wright and Hedley sailed for London from Blyth, and there are now in the Harbour to Load for London James Ogle, Clark's Ship. James Hannay, Edw.Robinson, Matt. Huntley and John Grant which will be dispatched as soon as possible. I make no doubt of Mr. T.Delavals vaunting of Success but you know the proof of the Pudding is in the eating, the Appearance is greatly on the other side

I am yours etc.

Matt. Ridley

P.S. No Ships Load with Byker this week for London

George Ward Esq   Newcastle, Sept 29 1775

I see by your favour of the 25th a great alteration in the Coal Market, it has not been usual to export such a quantity of Coals to America. There have sailed from Blyth to London since my last, Jas. Wood, Christ.Spanson, John Linton and Wm.Harrison, and remain in the Harbour to be dispatched to the same, James Hannay, Geo Brown, Robt Briggs, Richd Wright & Robt Steward, I hope they will be well received at Market.

(The following letter from the Duke of Northumberland confirms the evidence of the foregoing).

Northumberland House,
June 29 1771

Dear Sir,

I received the very obliging favour of your Letter and am very Sorry to find that there have been fresh Disturbances among the Colliers and Keelmen, as any Interruption of the Coal trade must be attended with great Inconveniences not only to the Neighbourhood of Newcastle but to Nation in general. However the prudent Measures which the Magistrates have taken will I hope put an effectual stop to them: And I do not doubt that the late Resolutions of the Council in order to encourage the Importation of Corn, will be a further Means of quieting the Minds of People and of restoring Peace and Industry. I had indulged Hopes that I should have paid my Compliments to you in the County before this time: but the Installation, which is fixed for the latter end of next Month will detain me till that ceremony is over.

The Duchess is arrived from abroad in good health and desires me to present her compliments. I am

Dear Sir,

Your most obedient Humble Servant,
Northumberland

M.Ridley Esq.

18. *Duties on coals sent to London, about 1803.*

Duties imposed on coals brought Coastwise
into London by London Chaldron

By Acts of Parliament		s.	d.	s.	d.
9 & 10 Willm. 3. c.13	..	5	0		
9 Ann (1710), c.22	..	3	0		
5 Geo. I. c.19 ⎱ Consolidated	..	0	10		
6 Geo.I. c.4 ⎰ May 1787	..	0	6	9	4

July 9 1797 ⎱
July 5 1803 ⎰ 5 per Cent

At Newcastle

Duke of Richmond	..		6		
Keel dues, light, &c.	..	1	9	2	3

47

London Dues

Metage ..	..	..	..	..	.. 0 4		
Orphan Duty ..	..	..	..	.. 0 6		
Additional Metage payable to							
the Orphans Fund ..	..	..	.. 0 4		
Coal Market ..	..	..	..	.. 0 1	1 3	

12 10

19. Extract from the case of the Northern Coal owners for the reduction of duties on coal, 1824.

1824. Statement of the Case of the Northern Coal Owners.

The Northern Coal Owners raise yearly a larger quantity of the best Coals than is required by the London and coasting markets, and in preparing them for the consumers are obliged to separate them from the small Coal, the greater part of which is unsaleable, on account of the heavy duties. This unsaleable Coal amounts in many collieries to one-fourth of the whole quantity worked, making a total within the two ports of four hundred thousand London chaldrons, which, if the existing duties were repealed, would be distributed around the coast, greatly to the advantage of the shipping, agricultural and manufacturing interests, and to the relief of the poorer classes of the maritime districts, who, under the present system, feel the pressure of taxation in the direct proportion of the difficulty they have in procuring an article essentially necessary for their health and comfort.

These duties upon Coals exported from the Tyne amount at the London market to 10s.8d., and at the coasting markets to 6s.6d.; and on Coals exported from the Wear, and Hartley and Blyth, to 10s.2d. at the former and 6s. at the latter per London chaldron.

The highest price charged by the Northern Coal Owners for Coals put on board is 16s.6d. per London chaldron, and the lowest for housekeepers. Coal is 8s. making the average upon different sorts less than 13s. per London chaldron; and any addition to this price to the consumer arises from circumstances over which the coal owners have no control, and in the advantages to be derived from which they do not participate.

20. *Memorandum on weights of Newcastle chaldron and canal ton, (about 1822).*

Weight of the Newcastle Chaldron and Canal Ton compared

The Canal Ton of Coals at the Pit is 20 Cwt. of 120 lbs each or 2520 lbs to the Ton. But they allow 168 lbs for waste on the passage to London and Call the Ton 2352 or 21 Cwt.

The Coals are sold at Newcastle by the Chaldron of 53 Cwt 112 lbs to the Cwt. Consequently the Ton of Coals at Newcastle only contains 2240 lbs being 1/8th or 280 lbs (less) than the Canal Ton.

The Newcastle Chaldron at the Pit		53 Cwt.
The Canal Ton	at ditto	22½Cwt.
Ditto	at London	21 Cwt.

No waste is allowed on the Newcastle Coals in the transport. Consequently we must compare them with the Canal Coals as delivered in London, that is as 53 Cwt the Newcastle or 28 Cwt the London Chaldron to 21 Cwt, to the Canal Ton.

The Newcastle Ten of Coals of 440 Bolls at the standard weight of 53 Cwt for 24 Bolls is 48·58 Tons. But as 21 to 22½ Bolls of the North Country Coals will weigh 53 Cwt instead of 24. The Ten will be 51·3 Tons. If we therefore state the Ten at 50 Tons it will be near the truth.

The Mine Rent to the proprietors of the Northern Collieries varies from 3^d. to 1^d/10 Ten.

Note:—So far as I can see the above calculation is not accurate for 20 cwt. of 120 lbs. amounts to 2400 lbs. and not 2520 lbs. as stated in the first paragraph. Either 21cwt. of 120 lbs. or 20 cwt. of 126 lbs. would give the figure stated. It would appear that the figure should be 21 cwt. of 120 lbs. at the pit, served by canals, and 21 cwt. of 112 lbs. at London the difference of 8 lbs. per cwt. giving a loss of 168 lbs. per canal ton.)

21. Coal exports from Newcastle and Blyth, 1798-1818.

An account of coals exported from Newcastle upon Tyne & Blythnook in each year from 1798 to 1818 both inclusive.

Year	Newcastle		Blyth	
	Coastwise	Foreign	Coastwise	Foreign
1798	395,985	44,756	37,979	166
1799	451,572	43,365	41,767	127
1800	537,793	47,487	43,336	104
1801	452,192	50,401	41,033	121
1802	494,488	44,000	44,821	20
1803	505,137	44,324	48,206	24
1804	579,929	52,589	44,518	138
1805	552,827	49,572	44,570	—
1806	588,277	46,683	49,413	195
1807	529,950	27,424	39,972	—
1808	619,125	16,001	48,603	960
1809	539,098	13,639	48,052	96
1810	632,299	17,253	47,330	44
1811	633,359	17,954	53,958	81
1812	630,633	24,985	55,258	—
1813	584,184	14,761	45,553	72
1814	649,151	31,986	48,529	—
1815	650,209	42,434	37,363	643
1816	678,151	43,783	49,417	771
1817	621,809	51,797	46,902	318
1818	671,871	47,685	51,397	441

22. Accounts for Blyth Salt-Pans, 1783-84.

1783	North Blyth Salt-Pans	£	s.	d.
	Dr.			
Jan	1 To Stock Per Balance for one Ton of Salt remaining on hand	2	0	0
	To Cash paid to:			
March 19	Thomas Freeman for cooperage to this day ..		7	0
May	3 Cuthbert Humphery for one years Poor cess ..		10	4

		£	s.	d.
May	12 William Hair for ¼ years Church cess	4	0	
	John Stobbs for ¼ years Land tax	6	10	
	The Salters for Shipping of Salt	5	3½	
	Elizabeth Richardson for Salters Smiths &c.			
	Allowance	1	7	3½
	Keel dues of Coals Sent up to No. B. Pans ..	2	0	10½
	The Salters their Wages to this day	16	3	0
	John Elliot for carpenters work	2	15	4
	Joseph Wheatley for Smith Work done this day	7	2	9
	Mark Weatherhead for Mason Work done do.	2	10	0
	Robert Dobson for conveying of Pan Rubbish	1	10	0
June	30 Jane Marshall for Raff etc. this Half year ..	11	1	2
Oct	26 Half a years land tax of the Pans..		6	10
Nov	22 Joseph Wheatley for Smith Work this Half ..	8	18	2
	John Elliot for Carpenter Work this Half yr. ..	1	3	6
	William Weatherhead for Mason Work ..	1	17	0
	Robert Robson for conveying of Pan Rubbish ..	1	10	0
Dec	31 The Salters their Wages to this day	33	5	6
	The Salters for Shipping of Salt, etc.	4	12	3½
	The Labourers Keel dues etc. to the Pans ..	8	1	7
	The Salters their Half Years Bounties	6	6	0
	Jane Marshall for Raff etc. this Half year ..	1	0	3
	John Lawson for a years Rent of the Pans ..	10	0	0
	For Pan Plates consumed this Year	10	0	0
	To Amount of Small coals consumed			
	being 595·⅓ Chaldrons @ 5s. per Chaldron ..	148	16	8
		284	1	8
	To Sir Matthew White Ridley Bart for Profits this			
	year	98	11	10
		382	13	6

1783   *Cr.*

By Cash received of:	Price (shillings)	Tons	Bushels	£	s.	d.
Jan 7 Sundry People for salt retailed	40	7	21	15	1	0
March 31 John Wood Esqr. for Salt Shipt	43	6	20	13	19	6
July 21 George Merryweather & Co. do.	40	40	0	80	0	0
Aug 1 Anthony Almond do.	38	20	0	38	0	0

			Price (shillings)	Tons	Bushels	£	s.	d.
Oct	14	George Merryweather & Co.	do. 36	10	0	18	0	0
Nov	4	Thomas Spence	do. 34	20	0	34	0	0
Dec	22	John Wood Esqr.	do. 34	6	0	10	4	0
				110	1	209	4	6
,,	31	Sundry People for Salt retailed this year	40	49	29	99	9	0
		By Stock of Salt remaining at the two Pans this day ..	32	40	0	64	0	0
		By Amount of 40 Chaldrons of Small Coals consumed by Smiths Salters and others @ 5s. per chaldron				10	0	0
		Total ..		199	30	328	13	6

1783		South Blyth Salt Pans			

			£	s.	d.
		Dr.			
Jan	1	To Stock Per Balance for Salt remaining at the Pans	100	0	0
		To Cash paid to:			
April	29	George Wigham for Half a years Land tax ..	1	15	7½
May	3	Elstob & Hogg for Half a year Poor cess ..	1	12	6
	12	Salters Wages to this day as per Cash Account..	49	12	10
		The Salters for Shipping Salt as per Cash Account	5	12	9½
		John Elliot for carpenters work done this Half Year	3	4	0
		George Brown & Jos. Wheatley for Smith Work do.	26	4	11½
		William Weatherhead for Mason Work done this ½ year	22	1	4
		John Watts for Smiths, Salters etc. Allowance do.	7	18	9
		Robert Dobson for conveying of Pan Rubbish do.	4	10	0
June	25	Jacob Bell for Tar for the Tubbs		7	8½
		Thomas Freeman for cooperage done do. ..		8	6
	26	The Salters their Half years Bounties of 6 Pans	9	9	0
		The Salters their Wages to this day	15	0	6
		Shipping of Salt etc. do.	2	5	3
	30	Mrs. Jane Marshall for Raff etc.	2	1	11

				£	s.	d.
Oct	28	George Wigham for Half years Land tax	..	1	15	7½
Nov	10	Robert Dobson for Half a years Poor cess	..	1	11	3
Dec	24	The Salters Half years Bounties of 6 Pans	..	9	9	0
		The Salters for Shipping Salt etc... ..	..	16	1	1½
		The Salters their Wages in full to this day	..	70	13	0
		Brown & Wheatley for Smith Work done	..	39	3	10
		John Elliot for Carpenter Work	..	1	17	2
		William Weatherhead for Mason Work	..	5	3	6
		John Watts for Smiths, Salters etc. Allowance	..	11	19	6
		Robert Dobson for conveying Pan Rubbish	..	4	5	0
	27	Joseph Turner for bricks and Pantiles ..	..	2	5	0
	31	Jane Marshall for Raff etc. this year ..	..	1	15	9
		To Amount of Small Coals consumed this year				
		being 1926½ Chaldrons @ 5s. per Chaldron	..	481	14	6
		To Amount of Panplates etc. consumed this year		22	6	4
				922	6	3
		To Sir Matthew White Ridley Bart for Profits				
		this year	..	411	13	3
				1333	19	6

Cr.

		By Cash received of:	Price (shillings)	Tons	Bushels	£	s.	d.
Jan	7	Sundry People for Salt retailed..	40	2	2	4	2	0
	15	George Merryweather & Co. for Salt Shipt	43	35	0	75	5	0
	28	Wake & Simpson for Salt Shipt him	43	10	0	21	10	0
March	8	Wake & Simpson	do. 43	10	0	21	10	0
	10	Geo. Merryweather & Co.	do. 43	39		83	17	0
	31	John Wood Esq.	do. 43	6	20	13	19	6
Apl.	10	John Tulf	do. 43	5		10	15	0
May	9	Thos. Spence & Co.	do. 42	20		42	0	0
	12	John Tulf	do. 42	6		12	12	0
	28	John Tulf	do. 40	12		24	0	0
June	3	Geo. Merryweather and Co.	do. 40	40		80	0	0
	29	Geo. Merryweather and Co.	do. 40	40		80	0	0
July	17	John Wood Esq.	do. 40	2		4	0	0
	22	John Wormald	do. 40	35		70	0	0
	24	Thos. Littlefair	do. 40	5		10	0	0

			Price (shillings)	Tons	Bushels	£	s.	d.
Aug	2 John Anderson	do.	40	6		12	0	0
	12 Geo. Merryweather	do.	38	36		68	8	0
	16 John Wood Esq.	do.	38	6		11	8	0
Sept.	15 Geo. Merryweather and Co.	do.	38	33		62	14	0
	23 Wake & Simpson	do.	38	10		19	0	0
	29 James Masterman	do.	38	10		19	0	0
Oct.	2 Jonan Sanders & Son ..	do.	38	5		9	10	0
	3 John Wood Esq	do.	38	8		15	4	0
	8 John Ward Esq.	do.	38	25		47	10	0
Oct.	14 Geo. Merryweather and Co.	do.	36	28		50	8	0
	25 Eliz. Richardson	do.	36	10		18	0	0
Nov.	3 John Tulf	do.	36	5		9	0	0
	4 Geo. Merryweather	do.	34	20		34	0	0
Dec.	6 Wake & Simpson	do.	34	10		17	0	0
	Jno Tulf	do.		1		1	14	0
	12 John Wells	do.	34	20		34	0	0
	20 Ann Smith and Co.	do.	34	33		56	2	0
	31 Sundry People for Salt retailed..		40	24	4	48	1	0
	Total			557	23	1086	9	6
	By Stock of Salt remaining at the Salt Pans this year			100	0	160	0	0
	By Amount of Small coals taken from the Pans by Keel etc. ..					87	10	0
	Total ..			657	23	1333	19	6

1784	North Blyth Salt Pans		£	s.	d.
	Dr.				
Jan	1 To Stock of Salt remaining Last year at the Sundry Pans being 40 tons @ 32s.		64	0	0
Dec	31 To Wages paid to this day per Cash Book		50	12	11½
	To Smith work	do.	18	16	0
	To Carpenter work	do.	5	9	0
	To Mason do.	do.	15	15	8
	To Keel dues do.	do.	13	4	5
	To Taxes & lesses	do.	1	16	0
	To Rent	do.	10	0	0

	£	s.	d.
To Leading Rubbish do.	3	0	0
To 559⅓ Chaldrons Small Coals @ 5s.	139	16	8
	322	10	8½
To Sir M. W. Ridley Bart., Profit this Year ..	29	18	11
	352	9	7½

Cr.   Price Tons Bushels £ s. d.
By Cash received of   (shillings)

Date		Price	Tons	Bushels	£	s.	d.
Aug	6 George Merryweather & Co. for Salt Shipt	32	27	0	43	4	0
Oct	18 George Merryweather & Co. for Salt do.	32	32	0	51	4	0
	Total Amount of Salt Shipt		59	0	94	8	0
Dec	31 Salt Retailed		57	8	114	8	0
	By Stock of Salt at No. Blyth Pans	32	83	22	133	13	7½
	By Coals used by Smiths Salters &c. .. .. ,, ..				10	0	0
			199	30	352	9	7½

1784   South Blyth Salt Pans

Dr.   £   s. d.

Date		£	s.	d.
Jan	1 To Stock remaining Last year at the Sundry Pans being 100 tons	160	0	0
Dec	31 To Cash paid Wages to this Day per Cash Book	129	10	5
	do. Allowances do.	2	18	1
	do. Smiths do.	95	15	8
	do. Taxes do.	6	12	5½
	do. Carpenter Work do.	53	17	6
	do. Masons do.	28	4	11
	do. Rent do.	25	0	0
	do. Leadings do.	8	10	0
	do. Geo. Huntley for Blood	1	10	0
	To 1856⅔ Chaldrons Small Coals @ 5s. ..	464	3	4
		976	2	4½

Cr.

By Cash received of:

	Price Tons (shillings)	Tons	Bushels	£	s.	d.
March 12 John Wood Esqr. for Salt Shipt him	34	6	0	10	4	0
22 John Wells of Boston for Salt Shipt	34	20	0	34	0	0
24 Capt. John Tulf for Salt shipt ..	34	5	0	8	10	0
April 3 Ann Smith & Co. for Salt Shipt	33	41	0	72	13	0
May 15 Isabel Richardson for do.	33	5	0	8	5	0
22 John Tulf for do.	—	1	0	1	13	0
26 John Cockerill for do.	33	26	0	42	18	0
June 16 John Wood Esqr. for do.	33	6	0	9	18	0
18 John Tulf for do.	33	5	0	8	5	0
July 22 Wake & Simpson for do.	33	12	0	19	16	0
Aug 6 Martin Mauser for old Panplates					16	0
30 John Anderson for Salt Shipt him	32	6	0	9	12	0
John Tulf for do.	32	2	0	3	4	0
Sept 9 John Wells for do.	32	20	0	32	0	0
10 Ann Smith & Co. for do.	32	37	0	59	4	0
13 George Merryweather for do.	32	31	0	49	12	0
20 John Tulf for do.	32	4	20	7	4	0
24 Thomas Wait for do.	32	30	0	48	0	0
Oct 18 John Tulf for do.	32	5	0	8	0	0
Nov 18 Wake & Simpson for do.	32	10	0	16	0	0
26 Ann Smith & Co. for do.	32	40	0	64	0	0
Total Amount of Salt Shipt ..		312	20	513	14	0
Dec 31 By Cash received of Sundry people for salt retailed	40	27	1	54	1	0
By Stock of Salt remaining at the Sundry Pans etc.	30	120	0	180	0	0
By Small Coals to Keelmen Smiths etc.				87	10	0
				835	5	0
By Loss this year to Sir M.W.Ridley's Account				140	17	4½
Tons		459	21	976	2	4½

56

(Note:—The large loss in this year appears to be due almost entirely to the amount spent on smiths', masons' and carpenters' work, and suggests either major repairs or the replacement of a salt pan. The total amount spent on these three items in 1784 is £177. 18. 1½d, in 1785 it was £67. 1. 0. and in 1786, £56. 1. 10½d. 1784 is the first year in which the payment for blood appears).

South Blyth Salt Panns from 22nd November 1790 to and with 22nd November 1791.

1791   *Dr.*   £ s. d.

	£	s.	d.
To Duty paid for 287½ Tons of Salt	2592	0	0
Paid Margaret Weatherhead & others on account of making Salt	107	15	6
do. for delivering Salt	17	14	2½
Sundry Salters, Ernest & Bounty Money	20	10	3
Mason Work	20	3	11
Smith Do.	33	14	3
Carpenter Do.	6	4	8
lesses & sundry other small disbursements	14	4	10¼
	2813	5	7¾

Due to sundries & placed to their respective accounts:—

	£	s.	d.
To Edward Watts for Salters drink Bills ..		10	0
To So.B.Houses, for rent of Salters Houses ..	7	0	0
To Colliery Account for 2359 Waggons of Coals @ 5S	589	15	0
To Mrs. Marshall, for Deals &c.	3	1	10
To Joseph Turner, for Bricks & Tyles	1	13	0
To Isaac Cookson Esqr. for Iron & Panplates	92	0	3½
	694	0	1½

		£	s.	d.
Totals		2813	5	7¾
		694	0	1½
To Profit & Loss, Gained		121	15	10¾
		3629	1	8

Cr.

			£	s.	d.
By Salt Duty as Per other side	..	.. ..	2592	17	11
By 347½ Tons of Salt Shipt	@ 30s. per B.		521	5	0
By 2535 Bushels sold by retail	@ 1s. per B.		126	15	0
By Thomas Spence, what was advanced him	..		1	2	0
By received, for one stamp	.. ..	..	0	0	6
By an abatement of one shilling per Waggon on 2359 Waggons the Quantity consumed		..	117	19	0
			3359	19	5
Balance in Stock		..	269	2	3
			3629	1	8

(Note:—The above account is particularly interesting for two reasons. Firstly, the payment to Margaret Weatherhead shows that women were employed on salters, a fact which is confirmed by the Earsdon Parish burial entries. Secondly, the abatement of one shilling per waggon was made by the Plessey Collieries, which were in the same ownership as the salt pans, and had it not been for this abatement the profits from the salt pans would have been only £3. 1. 10¾d.)

23. *Payments for Blyth Salt-Pans, 1774.*

1774		£	s.	d.
June	1 John Elliot's Bills from 22nd Nov. 1773 to 12 May			
	1774, For work at the Staith	5	5	9
	Eight Pans & Granaries	3	3	0
	Town & Lordship	2	17	0
	No. Blyth Pans	0	9	0
		11	15	6

Sept 17 Shipped in the *Young Uffrou Maria* of Stralsund, Charles Abraham Thuron, master, by order of Mr. Paul Jackson

		£	s.	d.

48 Tons Buckham's
38 Do. Hunter's
56 Do. Grey's
38 Do. Liddel's

	£	s.	d.
180 Tons of Salt 27s. per	243	0	0
Ballast & harbour dues	1	0	8
55 Matts @ 4d per	0	18	4
Messenger to Chirton	0	1	0
	245	0	0

Sent Invoice to M.R.Esqr.
N.B. £160 to M.R. & £85 to M.Gamul

Dec. 13 George Brown's half year's Accounts for smith work to 22nd Nov., viz

	£	s.	d.
Eight salt-pans	18	0	1
Staith £9. 16. 3.			
Do. per Keels 3. 3. 5.	12	19	8
Houses	3	13	0
	34	12	9

Dec. 13 Richard Wheatley's half year's Account to 22nd Nov., viz for the

	£	s.	d.
Eight Salt-pans	6	19	1
No. Blyth Pan	2	16	5
Taking the old pan in pieces	3	12	6
	13	8	0

Dec. 31 I weighed the Iron of the Old Pan taken to pieces at No.Blyth, the quantity was of

		£	s.	d.
Plates	23 cwts at 17s.	19	11	0
Do.	19 cwts at 8s.	7	12	0
Good Stuff	9 cwts at 26s.	11	14	0
9 Balks	31 cwts at 10s.	15	10	0
		54	7	0

24. *Analysis of Blyth salt prices, sales and profits, 1783-99*

South Blyth

Year	Wholesale Price per ton	Retail Price per ton	Quantity Sold Tons	Bushels	Profit £ s. d.	Loss £ s. d.
1783	43s. to 34s.	40s.	557	23	411 13 3	
1784	34s. to 32s.	40s.	339	21		140 17 4
1785	32s. to 28s.	40s.	474	6		23 3 9
1786	28s. to 26s.	40s.	248	33	23 3 0	
1787	32s. to 30s.	40s.	430	0	68 18 3	
		per bushel				
1789	28s.	1s.	202	18	2 18 0	
1790	28s. to 30s.	1s.	293	11	28 17 8	
1791	30s.	1s.	410	35	121 15 10	
1792	30s.	1s.	301	18	37 10 8	
1797	45s. to 50s.	2.s8d.	275	28	202 11 6	
1798	50s. to 55s.	1s.6d.	382	10	196 8 6	
1799	55s. to 60s.	1s.6d. to 2s.	225	19	217 15 10	
1799	55s. to 60s.	1s.6d. to 2s.	237	17	156 11 0	

North Blyth

Year	Wholesale Price per ton	Retail Price per ton	Quantity Sold Tons	Bushels	Profit £ s. d.	Loss £ s. d.
1783	43s. to 34s.	40s.	159	30	98 11 0	
1784	34s. to 32s.	40s.	116	8	29 18 11	
1785	31s. to 28s.	40s.	158	26	—	72 1 0
1786	26s.	40s.	122	14		16 0 4

(Note:—The account years vary as follows: 1783 to 1787 are from Jan. to Dec; 1789 is from Dec. to Nov; 1789 to 1792 are from Nov. to Nov; the 1797 to 1799 are from May to May, with the exception of the second 1799 which is from May to Dec. 1799.

25. *Letters and extracts from Matthew Ridley relating to the Salt-Trade, 1767-76.*

Mr. Thos. Williams   Newcastle, October 20th 1767
    Pudding Lane, London.

I received the favour of your letter and perceive that you had agreed for another Freight of Salt in Todrige, but you do not mention the

Quantity, only that he shall take in coals for dinnage, and that he provide proper Matts to keep the Salt clean, he is now at Blyth and the Owner Mr. Hannay says that the Ship with the Proper Dinnage will take 200 Tons, so that Quantity will be put on board the said Vessel and all imaginable care shall be taken to have the Salt shipped clean; I am sorry to have a complaint of the former cargo, the Salt Officers and my own people declare that they never saw better or cleaner Salt put on board any Ship, so probably the Damage may have been caused in unloading it; this parcel will be old and good, but if you send soon for another I wish that my salt at Blyth may be old enough but if you please to order the vessel to Shields I will deliver the Quantity there.

You shall hear from me with the Invoice when the Ship is despatched which will be by Friday's post, in the meantime I have given a Bill on you to Messrs. Airey and Carr for the Duty at 30 days from yesterday for £1140. 16. 8. which doubt not your duly honouring, I am

M.R.

Geo Ward Esquire   Newcastle, June 10th 1768

. . . . .I have now a large quantity of Salt there, and shou'd be glad to deliver 2 or 300 Tons in a month even at 26sh. per ton; money being at this time a very necessary Commodity; I shou'd be glad if you cou'd help me forward in this matter.

M.R.

Sir Geo. Colebrooke & Co.   Newcastle, June 29th 1768

. . . . .I am very well pleased to receive an order for Salt from your house, which is a Rarity, I wish that you cou'd find it agreeable to your interest and Inclination to renew our old Correspondence in that article, I have 500 Tons now at Blyth and if you will order 2 or 3 hundred forthwith; you shall have it at twenty six shillings per Ton, which is two shillings below what the Shields price is, with the additional advantage of being shipt free on board, which amounts to near a shilling more. I charge this parcel at 27sh. which I do by way of tempting you, having not sold any under 28sh. for some time. The Blyth coals are I think, at this time better than ever and hope the ships now loading there will by their cargoes prove it. I am with true regard

M.R.

61

Messrs. Peckover & Nelson   Newcastle, October 7th 1768

Since mine on the first returning your Bill on Messrs. Brown's &
Collinson £47. 5. 8. I have yours of the the first, Agreeable to your order
shall keep 20 Tons of Old Salt which shall be Ship'd whenever you send for
it, Salt must of course rise as it cannot be made at 28sh. and very little left
of Old Salt & not much made, a great many Salt Pans being pulled down
and converted to other uses, Bills in London are more convertable into
mony than in Lynn if the same to you.

J.G.

Mr. Wm. Tennis   Newcastle, August 4th 1769
at Clay near Holt, Norfolk.

Mr. Barrow sent me your favour of the 27th ultimo. The price of Salt
at Blyth is 28sh. a Ton and the best coals 13sh. per chaldron. The Duty
for the Salt must be paid in Cash or Bills in London at 30 days ere the
Salt can be Ship'd and the prime cost and Coals by Bills at 40 days when
Ship'd, the Salt is put on board at the above price clear of all charges.

I am etc.

J.G.

Geo Ward Esq.   Newcastle, 15 June 1770

. . . . The Salt Works at Blyth must be laid by in a Weeks
time if We have not a demand for Salt to make Room in the Garners, I
wish you cou'd sell for me 2 or 300 Tons at 26sh. some of the Salt is a
year and half Old, but for Reason above I must have a Quantity taken
away immediately.

Hannay had some expectation that Todridge wou'd be freighted for
that purpose. I hear Mr. Williams is Dead, pray who succeeds in the
Business. You will let me know their Firm, and how to draw for the
Duty and prime Cost if they shou'd agree.

Mr. Robt. Spencer   Newcastle, July 6th 1770

I have your letter of the 2nd and observe that Salt from Lymington
offered at 25sh. The difference in Quality between that Salt and Blyth
you are well acquainted with, however to encourage demand from Blyth,
I will deliver a Quantity here to be taken away forthwith at twenty-six
shillings per Ton and shall be glad to receive your orders.

George Ward Esqr.   Newcastle, Jan 4 1771

I am favoured with your letter of the 31st Decr. little or nothing has been done as to Dispatch in the Holidays, and the orders are so strict for pressing that the Keellmen are unwilling to move, but that I wou'd hope wou'd soon be corrected, if for the sake of London alone, the Coal Trade must not be stopt, I shall deliver to Mr. Bowles's Order two hundred Tons of Salt at 26sh. when the Ship arrives at Blyth, the charge of £1. 15. 9. Custom house ought to have been paid by the Master, I will get that matter settled with Mr. Couzens the Owner of the Vessel, and shall deduct that Sum out of the next Invoice. P.S. The Edward and Morton is despatched this Week with Byker 18 Keells Mr. Lamb.

Geo Hogg Esq. & Son   Newcastle, Jan 25 1771

I am favoured with your letter of the 21st and shall deliver on the Ships arrival, at Blyth 100 Tons of best Old Salt agreeable to your order at 30sh. free on board, but cannot engage for a further Quantity at the same price, it being now advanced I understand by the Dealers here to 35sh however when you favour me with Commands, you may depend on being served at as low a Rate as any person in this Country sells for. I am with true Regard

Mr. Wm Glass   Newcastle, June 28 1771

The price of Salt at Blyth is 35sh. a Ton put on board the Ship, which Mr. Meaburn desired I wou'd let you know by the post, The Salt is very good, For your Government the Duty is £5. 14. 1. per Ton and must be paid before put on board in Cash or London Bills at 30 Days the prime cost in Cash or bills at 40 days which when you please to favour me with your orders they shall be immediately complied with

(Note:—There is a mention of salt in the letters in September 1772 but a gap between that letter and the one of March 1776 which follows.)

George Ward Esq   Newcastle, March 15 1776

. . . . We are sadley let down in the Salt trade, all our Graneries being quiet full. I wish that you cou'd get some of the Contractors for the Victualling to order 2 or 300 Tons from Blyth they should have it 26sh per ton. I am Obliged to defer my Journey to London to see some of my Colliery affairs settled I am

Yours etc.

M. Ridley

Geo Ward Esq.   Newcastle, April 26 1776

I am favoured with your letter of the 22 with prices of Coals, if they wou'd keep up this, I think there wou'd not be reason to complain I mentioned in my last that Jubb was loading for London but his Owner Mr. Clark has now directed him to load for Lynn. When the Victualling Office contracted for salt to cure provisions they always used to specify what they called Castle Salt, as the Lymington and West County was not proper for curing provisions, There are not any Ships to report from Blyth for London this week, the whole Trade has been Coastwise.

I am

Yours etc.

M.Ridley

Geo Ward Esq.   Newcastle, June 14 1776

. . . . If you cou'd sell me Salt at 24sh. I wou'd consent, as the poor persons are out of employment.

Geo Ward Esq.   Newcastle, September 27 1776

. . . . .I will deliver at Blyth any Quantity of Salt even 500 Tons, as the Garners are full and the pans laid off; the present price is 26sh. per Ton but if that price cannot be had I will take 25 sh; I shall be much Obliged if you can get a Quantity disposed of.

I am

Yours

M.Ridley

26. Cost of building the New Brewery, 1784-86.

New Brewery at Blyth

	To Cash paid (for):	£	s.	d.
1784				
Aug	7 Boring for Water	1	9	2
1785				
July	8 Making a Well	3	6	6
	30 Stones for do.		10	0
	„ Laying in Lime Kiln		11	8

			£	s.	d.
Aug	6 John Brown & Co.	cleaning the Well	1	1	6
	22 Thomas Douglas	Stones	1	12	10
Sept	3 Geo. Brown	Smith work ..		11	10
	6 3 Keels of stones from Rocks		1	4	0
	17 Carpenter Work	per Note	1	1	0
	20 John Story	Chalk	2	11	6
Oct	1 John Elliot	per Note	1	10	0
	7 Wm. Graham	freight of slates etc.	8	8	0
	8 Thos. Maddison	20,000 Blue Slates	40	0	0
	,, Pilots Assisting to get in Timber			12	0
	15 John Elliot	per Note	2	6	5
	29 4 Keels of Stones from Rocks		1	12	0
Nov	5 John Marr	Hair per Note ..	11	4	3
	14 Geo. Brown	per Note	3	18	4
	26 2 Keels Stones & 1 Sloop Lime Stones		1	4	0
Dec	5 John Clark	for Ropes	1	2	2
	18 John Elliot	per Note	1	15	0
	,, John Watt	Limestones etc.	8	0	5
	24 John Elliot & Geo. Brown		1	0	8
1786					
Jan	7 John Elliot	per Note		10	0
Jan	,, Jane Marshall	Timber	5	11	9
	,, Joseph Turner	Bricks	35	9	6
Feb	6 Edwd. Charlton	Nails	16	4	8
	,, John Story	Beef	1	10	3
	,, 2 Keels of Stones			16	0
	18 Geo. Brown & Carpenters expenses		1	7	9
	,, Surtees & Co.	Timber & deals ..	30	12	6
March	4 James King & Co.	Fire Bricks ..	1	0	0
	,, John Elliot	per Note	1	9	6
	,, Geo. Brown	do.	1	5	0
	18 2 Keels flagstones from Rocks			16	0
	,, Geo. Brown	per Note		15	7
	,, Richd. Hetherington	Timber	102	7	3
	25 Surtees & Lambert	Deals	8	0	0
April	1 Geo. Brown		2	4	8
	,, 2 Keels Stones			16	0
	15 Geo. Brown		1	0	0
	29 do. & John Elliot		1	12	4
	,, Wm. Robinson	Note of Nails ..		8	8

			£	s.	d.
May	6 2 Keels of Stones ..	.. ..		16	0
	8 R. Hodgson & Co.	for Chalk	4	5	10
	12 James King & Co.	4,100 Blue Slates ..	9	4	6
	13 John Elliot	per Note	2	15	0
	„ Geo. Brown	do.		12	10
	27 do.	do.	2	6	10
	„ John Elliot	do.		15	0
June	3 Clearing Rubbish ..	.. ..		14	0
	10 Edwd. Watts	Allowance ..	6	3	0
	„ Geo. Brown	Smith		19	6
	24 Note of Mason Work	.. ..	8	4	1
	„ John Elliot	.. ..	1	5	10
	„ James King & Co.	Fire Bricks ..	3	0	0
July	7 Note of Mason Work	.. ..	6	0	5
	„ 1 Keel Stones ..	.. ..		8	0
July	7 Geo. Brown ..	.. ..	2	2	0
	„ John Elliot	.. ..	1	12	6
	22 Clearing Rubbish ..	.. ..		14	0
	„ Note of Mason Work	.. ..	5	3	1
	„ John Elliot	per Note	1	15	0
	„ Geo. Brown	do.		15	7
Aug	4 Gilbert Newton	Boring Pumps etc.	11	16	6
	„ Wm. Watson	Note of Plumber work	202	16	2
	5 Edwd. Charlton	Hoops & Nails ..	34	10	0
	„ Note of Mason Work	.. ..	5	0	6
	„ Geo. Brown	Smith		12	10
	19 Note of Mason Work	.. ..	4	13	4
	„ Geo. Brown ..	.. ..		8	1
	„ Henry Shadforth	Millstones	3	15	0
Sept	2 Wm. Spencer	Haircloth	1	10	4
	„ Note of Mason work	.. ..	4	1	9
	„ John Elliot & Gilb. Newton	.. .. ..	3	13	5
	„ Wm. Carr	Smithwork ..	22	6	3
	16 John Elliot	per Note		9	4
	26 Richd. Hall	for a Millstone ..	1	13	0
	31 John Elliot	per Note	2	5	4
	„ Wm. Carr	do.	1	5	1
	„ James Mills	do.	3	17	5
	„ Jane Marshall	Timber etc... ..	44	13	9
	„ John Rastrick	for Sundries ..	182	15	11

				£	s.	d.
Sept	31 Wm. Weatherhead	at Sundry times	..	239	4	8
	„ Wm. Carr	Smith	.. ..	7	7	0
Oct	14 John Elliot & Gilbt. Newton	.. ..	..	1	6	0
	31 Geo. Emerson & Co.	.. .. ..	..	10	11	2
	„ John Marr	Hair..	..	2	9	3
	„ Wm. Carr	Smith work	..		15	8
	„ James Mills	Mason	..		19	4
Nov	25 do.	do. ..	..	1	6	0
	„ Note of Mason Work at Cellar	.. ..	..	2	15	5
	„ James Archbold	Slater	..	61	19	0
	„ John Patterson	Locks & Snooks	..	2	17	6
	„ Edwd. Watts	Allowance	..	6	4	4
	„ John Elliot	Agreement ..	..	48	3	4
	„ Joseph Turner	Bricks etc.	..	6	12	0
	„ Jane Marshall	Timber etc...	..		—	
	„ John Marshall	Ropes	..	1	0	6
	„ Robt. Moor	Painting etc.	..	25	8	4
	„ Geo. Brown	flywheels etc.	..	6	4	8
	„ Edwd. Poad	freight etc. ..	..	6	2	2
	„ do.	do. ..	..	3	17	8
	„ Wilkinson & Co.	Iron	..	3	15	4
	„ Edwd. Robinson	Slaters Allowance		5	5	4
				1340	13	9

27. *Analysis of costs in building the New Brewery, 1784-86.*

					£	s.	d.
Mason	..	..	..	..	.. 281	6	0
Plumber	..	..	..	..	.. 202	16	2
Timber	..	..	..	..	.. 191	17	3
Engineer		..	..	..	.. 182	15	11
Carpenter		..	..	..	.. 74	10	10
Slater	..		..	..	.. 61	19	0
Slates (24,100)		..	..	..	.. 60	14	7
Bricks	..	..	..	..	.. 42	1	6
Hoops & Nails	..	..	..	..	.. 41	3	4
Smith Work		..	..	..	.. 26	0	1

					£	s.	d.
Painting	..	..	..	..	25	8	4
Hair	..	..	..	..	15	3	10
Lime	..	..	..	..	14	17	3
Boring Pumps	..	..	..	..	11	16	6
Freight	..	..	..	..	9	19	10
Stones	..	..	..	..	7	12	10
Flywheels	..	..	..	..	6	4	8
Millstones	..	..	..	..	5	8	0
Fire Bricks	..	..	..	..	4	0	0
Iron	..	..	..	..	3	15	4
Locks	..	..	..	..	2	17	6
Ropes	..	..	..	..	2	2	8
Clearing Rubbish		..	..	..	1	8	0

28. *Accounts for grains, malt and hops, 1774.*

1774   £   s.   d.

May   3 Grains sold from Xmas to and with the 3rd instant, viz to

				£	s.	d.
R. Shotton	319 B. at 1½d.	..	..	1	19	11
R. Dobson	319 B. at Do.	..	..	1	19	11
Elstob & Hogg	319 B. at Do.	..	..	1	19	11
				5	19	9

Cr. Brewery

Dec   16 Hops Bought of Mr. Ogle Wallis, viz

               Cwt. qtr. lb.

	Cwt.	qtr.	lb.		£	s.	d.
Sept 5 Two Pockets	2	3	11	at £5	14	4	10
Oct 8	2	3	14	at £4.6	12	7	3
Nov 11	26	0	23	at £4.8	115	6	0
do. New Hops	5	0	19	at £4.2	21	4	0
					163	2	1

Cr. Malting

Dec   31 Malt delivered on trust since Midsummer viz, to

bushels

						£	s.	d.
Brewery	1623	at 4/8 per	..	..		378	14	0
M. R. Esqr.	225	5/0 per	..	..		56	5	0
V. Pearson	18	Do.	..	..		4	10	0
J. Robinson	12	Do.	..	..		3	0	0
Wm. Harrison	10		..	..	..	2	10	0
Thos. Gibson	15	..	..	..	..	3	15	0
E. Hannay	8	..	..	..	..	2	0	0
R. Humphrey	6	..	..	..	..	1	10	0
H. Elliot	1	..	..	..	..	0	5	0
						452	9	0

Cr. Malting

29. Blyth Brewery Accounts, 1788-89.

Blyth Brewery from 31st of Decr. 1788 to & with 22nd Novr. 1789

Cr.	£	s.	d.	£	s.	d.
To Value of the Brewery & Moveable Stock						
at 31st Decr. 1788				2364	10	9
Paid Sundry's Wages from 31st Decr. 88 to						
& with 22 Novr. 89	105	10	7			
James Mills for Mason Work	1	10	6			
John Elliot for Joiner Work	1	19	11			
Drayman's allowance & Sundry expenses						
& returns	5	8	6			
Sundry Taxes & lesses	18	0	6			
Thos. Turner for Cooperage	2	16	7	135	6	7
Sundries for Barley				219	7	10
Mr. Jackson Duty on Ale & Malt ..				504	7	6½
Messrs. Coffin & Sons for Hops				125	13	6
Newsham Estate for Rent of Ground ..	25	0	0			
Sundry small disbursements	10	8	3	35	8	3
				3384	14	5½

Due to Sundries 22nd Novr. 1789

	£	s.	d.	£	s.	d.
To So. Blyth Houses, for Rent of Workmen's Houses	8	0	0			
To Colliery, for Coals & Cinders	28	7	6			
To Mr. Ogle Wallis, for Hops	31	19	6			
To Mr. Jackson, for Duty on Ale now due	53	6	10			
To Edwd. Charleton, for Nails	3	5	1½			
To Robt. Moor, for Sundries	7	5	0			
To Robt. Stoker for 6 Half Barrells ..	2	2	0			
To Jno. Marshall, for Sundries	1	8	7			
To Mrs. Marshall, for Do.		14	8	136	9	2
To Profit & Loss, Gained				274	9	10
				£3795	13	6

Cr.

£ s. d.   £ s. d.

Sold & delivered from Brewery from
31st Decr. 1788 to 22 Novr. 1789:

		£	s.	d.	£	s.	d.
2115¼ Half Barrells of Ale	at 16/- per	1692	4	0			
95 Do. of Ale & small beer	at 6/- per	28	10	0			
2½ Do. of Do.	at 5/- per		12	6			
546½ Do. of Small Beer	at 4/- per	109	6	0			
47 Do. of Do.	at 3/6 per	8	4	6			
					1838	17	0
Recd of James Matthewson for yeast & small Beer by retail					28	5	6
„ of Sundries for 3952 Bushells of Grains at 2d. per					32	18	8
Sir Matthw. W. Ridley got at Blagdon							
507 Bushells of Malt	at 6s. per	152	2	0			
394 lbs. of Hops	at 18d. per	29	11	0	181	13	0
Recd of John Clark for Hops						9	4
					2082	3	6
Balance in Stock & Value of the Brewery					1713	10	0
					£3795	13	6

30. *Duties on Beer, Malt and Hops.*

	Date	Description	Duty s. d.	
Beer	1674	Strong beer exceeding 6s. per barrel ..	1 3	per barrel
		Small beer 6s. or less ,, ..	3	,,
	1705	Strong beer as above ,, ..	2 0	,,
		Small beer as above	6	,,
	1709	Strong beer as above	2 3	,,
		Small beer as above	7	,,
	1710	Strong beer as above	5 0	,,
		Small beer as above	1 4	,,
	1761	Strong beer as above	8 0	,,
		Small beer as above	1 4	,,
	1782	Strong beer above 11s. per barrel	8 0	,,
		Table beer above 6s. but not above 11s. per barrel	3 0	,,
		Small beer 6s. and under per barrel ..	1 4	,,
	Duty yield in 1767	£538,000	
	,, ,, 1774	£1,358,400	

Malt	1697	Malt made in England and Wales	6	per bushel
	1760	,, ,, England	9¼	,,
		,, ,, Scotland	4½	,,
	1780	,, ,, England	1 4½	,,
		,, ,, Scotland	8	,,
	1785	,, ,, Ireland	7	,,
	1791	,, ,, England	1 7½	,,
	1792	,, ,, ,,	1 4½	,,
	Duty yield in 1760	£963,000	
	,, ,, 1792	£1,203,000	

Hops	1710	All hops grown in Great Britain	1	per lb.
	1779	,, ,, ,, ,,	5%	increase
	1781	,, ,, ,, ,,	do.	,,
	1782	,, ,, ,, ,,	do.	,,
	Duty yield in 1774	£138,800	
	,, ,, 1792	£151,000	

31. *Leases of the Blyth Ropewalks, 1762-1813.*

Leases to the Clark family.

On 26 January 1762 Sir Matthew White Ridley leased to John Clark for 21 years a parcel of ground 400 yards long and 6 yards broad for a ropework at a yearly rent of £2.

On 1 August 1770 the rent of the ropework was fixed at £3 per annum and John Clark was granted a lease of a piece of ground 20 yards long and 7 yards broad on which to build conveniences which he thinks proper.

On 1 May 1784 he leased a house at South Blyth for a period of 15 years at a yearly rent of £17.

On 20 November 1795 he leased the house and ropework on pieces of ground 400 by 6 and 20 by 7 yards respectively, and also a piece of ground adjoining the ropework which measured 30 by 7 yards, together with the yarn house and tar house and also all other erections and buildings erected and built upon a piece of ground adjoining the said ropework on the west side, all of which premises were stated to be in the occupation of John Clark and of William Scrogs. The term of the lease was for three lives and the yearly rent was £3.

In 1810 the lease was again renewed for three lives on the same terms to Charles Taylor Clark, son of John Clark.

Leases to the Marshall family.

On 23 October 1762 George Marshall leased part of a field in the occupation of Roger Shottom measuring 50 yards by 30 yards for a term of three lives of George Marshall, and George Marshall and Jane Marshall (his son and daughter). By his will dated 19 March 1773 and proved on 10 September 1774 George Marshall left his mansion house consisting of four rooms on a floor to John Marshall, his son, with a life interest to Jane Marshall, his wife. A list of leases includes: "Oct 30 1770 Lease of Ropery to Marshall at Blyth for 3 lives.

On 11 November 1782 John Marshall leased a ropework 400 by 6 yards for 15 years at a yearly rent of £3, with a right to renew on payment of £15.

On December 1799 this lease was renewed as from 12 November 1797 for 15 years at £3 in the names of John and Mark Marshall, and a further renewal was made on October 1813 on the same terms.

32. *Brickmaking accounts, 1774.*

1774					£	s.	d.
Dec	23 Joseph Turner has burnt this year						
	126400 Bricks at 12d. per	..	..	..	6	6	4
	87200 Do. at 11s. per	..	..	..	47	19	2
			Rent	..	54	5	6

1774

Dec 23 Bought of Joseph Turner this year, viz. for

	£	s.	d.			
Town & Lordship						
18700 Bricks 12d. per ..	11	4	4½	15	6	4½
2050 Tiles 40s. per	4	2	0			

Plessey						
34000 Bricks , , ..	20	8	0			
5500 Pantiles	11	0	0			
80 Ridge Tiles	0	6	8			
				31	14	8

Eight Pans						
3400 Bricks	2	0	9			
400 P. Tiles	0	16	0			
				2	16	9

No. B. Pans						
400 Bricks	0	4	9			
200 Pantiles	0	8	0			
				0	12	9

Blagdon						
800 Pantiles	1	12	0			
600 Bricks	0	7	2	1	19	2
				52	9	8½

F

33. *Blyth-built ships registered in the Ports of Newcastle upon Tyne and Whitby, 1786-99.*

Date of Regn.	Name of Ship	Description	Tonnage	Owner	Year built and builder
3/3/87	Prosperous	Pink sterned Sloop	34	Thos. Wardall	1757
20/10/86	George & Jane	Pink std. Pink	228	Jane Marshall	1761
9/10/86	James & Mary (1)	Sq. std. Snow	184	Robt. Stoker	1764
13/10/86	Sheepwash	Sq. std. Sloop	50	Ed. Watts	1775
11/11/86	Ranger	Sq. std. Barque	254	Robt. Cowerd Hy. Cowerd	1782
7/12/86	Adelphi	Sq. std. Snow	231	Edmund Hannay	1782
18/9/86	Nautilus	Sq. std. Brigantine	216	Ed. Hannay	1783
21/9/86	John	Sq. std. Sloop	53	Henry Smart Robert Smith John Errington	1783
21/9/86	Ridley	Sq. std. Barque	201	Richard Hetherington Wm. Gray & Robert Smith	1783
16/10/86	May Flower (2)	Sq. Std. Brigantine	101	John Storey John Watts	1784
16/5/1801	Marys	Sq. Std. Brigantine	258	Ed. Cook Millburn John Hallowell John Robinson	1784
13/9/86	George & Jane	Sq. Std. Brigantine	175	Geo. Emmerson	1785
18/9/86	Diamond	Sq. Std. Sloop	43	Edmund Hannay	1785
21/9/86	John	Sq. Std. Sloop	41	John Clark John Annett	1786
23/9/86	John	Sq. Std. Sloop	52	Thos. Thrift Jasper Browell John Crooks Alice Wood Elizth. Dobson	1785
30/9/86	Swan	Sq. Std. Sloop	41	Ed. Watts George Gowan	1785
28/9/86	Dorothy	Sq. Std. Sloop	68	George Potts Philip Cowell	1786
12/9/86	Hope	Sq. Std. Snow	212	Edmund Hannay	1786
17/8/87	John	Sq. Std. Sloop	69	Edmund Hannay	1786
10/6/87	Elizabeth	Sq. Std. Sloop	45	John Clark	1787 M. Watson

74

10/5/88	Crofton	Sq. Std. Sloop	47	Robert Stoker	1788
8/9/88	Gemini	Sq. Std. Brigantine	171	John Clark	1788
					M. Watson
13/10/89	Chancellors	Sq. Std. Brigantine	204	Edmund Hannay	1789
					E. Hannay
3/9/90	Lord President	Sq. Std. Brigantine	98	Edmund Hannay	1790
					E. Hannay
19/4/92	Neptune	Sq. Std. Snow	278	Edward Watts	1792
					E. Watts
30/4/94	Westmorland	Sq. Std. Brigantine	257	John Watts	1794
				Matthew Wilson	E. Watts
18/5/95	Young Edward	Sq. Std. Brigantine	111	Edward Watts	1795
					E. Watts
5/7/96	Perseverence	Sq. Std. Brigantine	161	Edmund Habnay	1796
					E. Hannay
30/6/97	Edward & Mary	Sq. Std. Brigantine	184	Edward Watts	1797
					E. Watts
3/6/97	Ruby	Sq. Std. Slocp	48	Edmund Hannay	1797
					E. Hannay
9/2/98	Ridley	Sq. Std. Brigantine	78	Robert Stoker	1797
					R. Stoker
9/10/98	Edmund	Sq. Std. Brigantine	131	Edward Watts	1798
					E. Watts
25/11/98	Ceres	Sq. Std. Brigantine	107	Edward Watts	1798
					E. Watts
1/6/99	Perseverence	Sq. Std. Brigantine	96	John Robinson	1799
				Geo. Davison	R. Stoker
19/11/99	Lady Tyrconnel	Sq. Std. Brigantine	101	Edward Watts	1799
					E. Watts

Blyth-built ships registered in Whitby, 1786

1786	Constant Ann	Sq. Std. Sloop	44		1740
1786	Thomas & Jane	Sq. Std. Sloop	46		1774
1793	Blyth	Sq. Std. Sloop	55		1793

Blyth-built ships recorded in Lloyds Register, 1786

1786	William & Margaret	Brig	140		1776
1786	Betsy	Brig	80		1784
1786	Integrity	Brig	150		1784

1. Re-registered on 12/9/98, with this note: "Registered at this port the 9th October 1786, the certificate whereof no. 222 was left by the master on board the said ship when captured by the enemy".

2. Re-registered on 21/4/90 after lengthening and increase of tonnage to 140 tons.

34. *Number of ships built in Blyth, 1750-1799.*

Year	Number	Year	Number	Year	Number
1750	1	1783	3	1792	1
1757	1	1784	4	1793	1
1761	1	1785	4	1794	1
1764	1	1786	4	1795	1
1774	1	1787	1	1796	1
1775	1	1788	2	1797	3
1776	1	1789	1	1798	2
1782	2	1790	1	1799	2
				Total	41

35. *Ships employed in the Blyth coal trade, 1790-99.*

(Blyth-built ships marked with asterisk)

Name of Ship	Years working	Total cargoes	Name of Ship	Years working	Total cargoes
Ann & Margaret	1797	1	Goldthorpe	1793-8	43
Boreas	1797-9	4	Hesperus	1793-7	8
Caledonia	1790-2	6	Holderness	1791-6	3
Ceres*	1790-9	36	Hope*	1790-9	70
Chancellors*	1790-6	47	John & Mary	1793	2
Charming Sally	1790-4	27	John*	1790-7	50
Claude	1796-8	10	John & Betsy	1790-9	49
Crescent	1791-2	3	James & Mary	1790-8	41
Crofton*	1793-5	19	Lady Ridley	1799	1
Defiance	1790	1	Leviathan	1797-9	4
Dorothy*	1791-3	18	Lord President*	1790-1	3
Eagle	1793-8	40	Mary & Jane	1794-7	11
Ebenezer	1793	1	Marys	1793	1
Edmund*	1798-9	2	Mayflower*	1790-9	60
Edward & Mary	1790-8	24	Mediator	1796-9	16
Fanny	1790-1	13	Mercury	1790-9	43
Fortune	1790-9	75	Minerva	1798-9	10
Friendship goodwill	1798-9	7	Nautilus (big)*	1791-9	53
Gemini*	1790-6	29	Nautilus (small)*	1798-9	12
George & Jane*	1790-5	13	Neptune*	1792	1

Name of Ship	Years working	Total cargoes	Name of Ship	Years working	Total cargoes
Peggy	1798	1	Robert	1791-4	9
Perseverence*	1796-8	21	Robert & Margaret	1790-9	37
Polly	1790-6	27	Sally	1798	1
Providence	1797-8	10	Speedwell	1798	5
Rachel	1791-9	58	Thomas & Alice	1790-6	33
Renown	1790-6	52	Three Brothers	1797-9	8
Ridley*	1798	6	William	1794-6	16
			Young Edward*	1795	15

36. *House rents due in Blyth 11th Novemr 1787.*

	Folio	Due £	s.	d.	Paid £	s.	d.
Custom House	1	4	—	—			
Amor. Pattison	2	6	—	—			
Jonas Polwart	3	2	—	—			
Wm. Mark Watson	6	21	—	—			
Edwd. Potts	8		10	—			
Thomas Gleghorn	9	1	—	—			
Ann Roberts	11	2	—	—	1	—	—
John Alexander	14	3	10	—			
Wm. Pollard	17	2	8	—			
Wm. Miller	18	5	5	—*			
Eliz Smith	19	2	11	—*			
William Watson	23	13	15	8*			
Henry Taylor	24	1	10	—			
William Hague	26	1	2	6			
David Davison	27	2	2	6			
Thomas English	28	60	—	—*			
Geo. Storey, Waggon Wright	31	0	2	6			
Goe. Tate	32		10	6	10	6	
William Oswald	34	2	10	—			
Timothy Davison	36	15	8	—*			
Gilbert Newton	37	5	0	—*			
Richard Wright	41	7	—	—			
John Jubb	43	2	0				

					Folio	Due £ s. d.			Paid £ s. d.		
William Carr	..	..	..	..	44	1	10	—			
John Hudson	..	..	..	..	45	2	19	6*			
Robt. Brown	..	..	..	..	46	1	—	—			
James Laws ..	..	..	..	..	47	9	—	—*			
John Watts, Senr.		..	..	..	50	9	0	—			
Elizth. Wilson	..	..	..	..	53	3	10	5			
James Clinton	..	..	..	..	54	27	0	—*			
Geo. Hall ..	..	..	..	..	55	3	18	—*			
Isabella Robson	..	..	..	..	58	2	0	—			
Adam Ferguson	..	..	..	..	60	1	10	—			
Robert Shanks	..	..	..	..	61	3	10	—	3	0	—
Edw. Robinson	..	..	..	..	63	12	0	—			
Robert Gibson	..	..	..	..	65	13	19	—*			
Richd Gibson	..	..	..	..	66	4	10	—			
William Hetton	..	..	..	..	67	16	0	—			
William Atkins	..	..	..	..	68	2	18	—*			
Thomas Smart	..	..	..	..	70	2	5	—*			
James Henderson	..	..	..	..	71	3	—	—*			
John Thompson	..	..	..	..	73	3	—	—*			
Charles Wilson	..	..	..	..	74	1	—	—			
Robt Mitchell	..	..	..	..	75		10	—			
Ann Nesbit ..	..	..	..	..	76	3	0	—*			
William Brockett	..	..	..	..	77	1	—	—			
Thomas Wright	..	..	..	..	78	1	0	—			
Mary Humble	..	..	..	..	79	1	—	—			
Ann Sibbett	..	..	..	..	80	1	10	—			
James Nicholson	..	..	..	..	81	1	10	—			
Jane Watson	..	..	..	..	82	1	10	—		10	—
Barbara Potts	..	..	..	..	83	3	—	—*			
Elizth. Gray ..	..	..	..	..	84	6	10	6*			
Isabella Thompson	..	..	..	..	85	2	0	—*			
Don Richardson	..	..	..	..	88	2	10	—			
Ann Mills ..	..	..	..	..	91	1	15	—			
Wm. Oliver ..	..	..	..	..	92		10	—			
Dorothy Nicholson ..	..	..	..	94	10	0	6			
John Lishman	..	..	..	..	95	3	3	6			
Margt. Crow ..	..	..	..	..	96	5	10	—*			
Robt. Maflen	..	..	..	..	98	1	0	—			
Geo. Shanks	..	.. ° °	..	..	99	10	13	6			
Thomas Mather	..	.. ˮ ˮ	..	..	99	3	—	—*			

	Folio	Due £	s.	d.	Paid £	s.	d.
Jos. Rogers	100	1	10	—			
Rose Jubb	101	1	10	—			
James Lamb	102	2	0	—			
Eliz Wilson	104	15	0	—*			
Abm. Elstob, Butcher	105		15	—			
Geo. Huntley	106	4	2	6			
Widd. Briggs	107		5	—			
Robert Briggs	108	1	10	—			
Margt. Brown	109		10	—			
Cuth. Fouth	110	1	2	6			
Geo. Ditchen	111		2	6	0	2	6
Willm. Fairbairn	113	1	0	—			
Wm. Stoker	114	1	10	—			
Jos. Hopper	115		15	—*			
Thos. Kennedy	116		15	—			
Eliz. Farm Whitley	118	2	0	—			
William Twizell	121	7	0	—	7	0	0
James Collins	122	3	10	—*			
John Smith	123	1	—	—			
William Hedley, Junr.	124	3	19	—*			
John Hedley, Fishman	125	5	10	—*		15	—
Richard Twizell, Pilot	126		8	6			
Margt. Twizell	127	1	10	—			
John Easterby	128	1	—	—			
Richd. Short, Senr.	129	10	4	—*			
John Short, son of do.	130	7	2	6*			
Richard Short, junr., sailor	131		15	—			
Thomas Redford	132		15	—			
John Short, son of Geo. Short	133	1	10	—			
Robert Hedley, Pilot	134	1	10	—			
John Sharp	135		10	—			
John Twizell	136	2	17	—*			
Eliz. Gutterson	137	5	19	—*			
Robert Freeman	138	1	5	—			
Thomas Wite	141	13	10	—			
William Blakey	143	1	0	—	1	—	—
Robt. Turner	144	4	0	—			
Geo. Shanks, Stable	145	1	1	—			
William Burn, senr.	146		15	—			
Geo. Brown, mariner	149	4	0	—	4	—	—

	Folio	Due			Paid		
		£	s.	d.	£	s.	d.
John Thompson, sailor	150	3	9	—*			
James Nichols, junr,. do.	152	1	0	—			
Ebenr. Kell	153	1	0	—			
John Campbell, sailor	154	1	—	—			
Jno. Manners..	155	2	0	—	2	—	—
Margt. Kirkup	158		15	—		10	—
William Burn, junr.	160		10	—			
Henry Finley	159		5	—			
		470	16	1			

Those marked thus * are suposed doubtfull & bad Debts.

R. Hetherington.

37. *Leases of land at South-Blyth, 1759-1800.*

23rd April 1760. To Elizabeth Huntley:—all that messuage burgage or tenement with the appurtenances and garth or yard behind the same containing in length from the great stable leading to the waggon way sixteen yards and in breadth eleven yards and the said garth or yard behind the same containing in length sixteen yards and in breadth five yards. Rent, 10s. For term of 21 years from 1st May 1759.

23rd. October 1762. To George Marshall: part of a field in the occupation of Roger Shotton extending from the south corner of the garden wall in a direct line southwards 50 yards and in breadth 30 yards.
(Note: The property built on this site became the Ridley Arms).

1st August 1763. To George Stephenson of Newcastle upon Tyne; a piece or parcel of ground lying and being at South Blyth extending from the Custom House in a direct line length 54 feet from east to west 54 feet. Rent, 10s. For term of three lives.

30th October 1765. To Mrs. Elizabeth Wilson: a messuage in South Blyth and stable 12½ yards in length and 6 yards 2 feet in breadth. Rent, £8. For 21 years.

9th December 1766. To Henry Wilson: a piece of ground at South Blyth with liberty to build a house thereupon 10 yards square adjoining a messuage or tenement known by the name Cansfields House. Rent, 5s. For 21 years.

22nd May 1770. Notice of Sale. To be sold to the highest bidder all that new built messuage or tenement situate at South Blyth aforesaid now in the possession of the said Mrs. Hunter and of Wm. Oswell as tenants thereof at and under the yearly rent of £22 and all other the premises now held under leave for three lives from the late Matthew Ridley Esqr.

(Note: John Young was the highest bidder at the sum of £340.)

11th November 1772. To Captain Francis Wright, master mariner: that parcel of ground containing in quantity 22 yards square and situate lying and being in the town of South Blyth aforesaid on the east side of the main street there whereon the said Francis Wright with the permission of the said Matthew Ridley hath lately erected and built a brick house and shop for and during the natural lives of Edward Wright son of the said Francis Wright, William Clarke son of William Clark of South Blyth aforesaid Miller and John Story son of William Story of the same place Butcher, Rent, 20s. For three lives.

Note:—This property was conveyed to George Wright of Seaton Sluice, Officer of the Customs executor and devisee of Francis Wright on 21st Novr. 1811.

2nd February 1773: To Hannah Greeves: a large brick house at South Blyth at the North End. Rent, £5. For 9 years.

31st December 1774: To Jno. Boulby: land measuring 50 by 30 yards occupied by Geo. Marshall. Rent, 30s. For three lives.
(This appears to be the land on which the building left under the will of George Marshall, proved on 10th Sept. 1774, stood which became the Ridley Arms.)

30th January 1776. To Thomas Douglas of Bedlington: a piece of land at South Blyth length 22 yards breadth 22 yards boundering on a messuage or tenement belonging to Mr. Francis Wright on or towards the south on the Kings High Street. Rent, 21s. For three lives.

1st June 1777. To John Brotherick of Hartley Pans, gentleman, and George Tate of the same place, glazier; a parcel of waste ground at South Blyth, adjoining the south or south east end of house belonging to Mr. Francis Wright for 11 yards and to the north or northeast 44 yards. Rent, 30s. For three lives.

6th August 1777. The land conveyed to Thomas Douglas: on 30 Jan. 1776 was conveyed to George Dormand of Newburn for a year at one peppercorn. Another lease for a year was signed on 12th April 1779, and on 27th September 1780 it was conveyed to Edward Lawson.

22 November 1782. To George Detchon of Newsham, weaver: all that piece or parcell of ground whereon a cottage or dwelling house hath been lately erected and built situate lying and beingat Red House in the said parish of Earsdon containing forty yards in length and fourteen yards in breadth as the same marked and set out adjoining Blyth Waggon way. Rent 5s. For three lives.

22nd November 1782. To Cuthbert Forster of South Blyth, carpenter, all that piece or parcell of ground whereon two cottages or dwelling houses have been lately built situate lying and being at Red House in the said parish of Earsdon containing fourteen yards in length and eighteen yards in breadth as the same is now marked and set out adjoining the old salt pans on the waggon way near South Blyth aforesaid and at the south end of a certain field called the Pittfield and west of the ropework in the possession of Mr. John Clark. Rent, 25s. For three lives.

1st May 1784 to Joseph Ramsay a messuage at South Blyth consisting of a dwelling house stable and outhouse. Rent, £12. 12. 0 For 15 years.

1st May 1784. To John Clarke: a messuage at South Blyth. Rent, £17. For 15 years.

19th January 1785. To William Burn of South Blyth, ropemaker, and Ann his wife: all that messuage tenement or dwelling house lately built and erected by the said William Burn upon the freehold of the said Sir Matthew White Ridley situate and being at South Blyth aforesaid at a place commonly known by the name of the Folly Pans now in the tenure or occupation of the said William Burn and of Matthew Lindsay his tenant. Rent, 15s. For 3 lives.

(Note:—The above conveyance read in conjunction with those of Novr. 1782 to Detchon and Forster appears to place the three houses at the bottom of field number 64 and the Folly Pans on the Slake where it was crossed by the waggon way. The house at the junction of Park Road and Plessey Road was at one time known as the Folly.)

1st April 1790. To Matthew Wilson, mariner: land measuring 15 yards by 9 yards 2 feet and a house consisting of four rooms in the occupation of Matthew Wilson and Elizabeth Wilson his mother. Rent, 2/6 a room. For 3 lives.

31st October 1797. To Edward Twizell, master mariner: that piece or parcell of ground situate and being in Blyth in the county of Northumberland containing in length from north to south twenty six FEET or thereabouts and in breadth from west to east seventy five feet or thereabouts and which said piece or parcell of ground is bounded on the Town Street of Blyth on the west, the River Blyth on the east a messuage or dwelling house belonging to Sir Matthew White Ridley on the south and a piece of waste ground belonging to the said Sir Matthew White Ridley on the north. Rent, 5s. For 42 years.

26th January 1799. To John Charlton, grocer: a piece of ground 48 ft 6 inches in length and 36 feet 6 inches in breadth boundering on premises in the occupation of George Story, butcher, on the east west and north and on the Waggon Way of the said Sir Matthew White Ridley on the south. Rent, 5s. for 21 years and 10s. for 21 years, the total term being 42 years.

(Note:—This site would appear to be in the corner of field No. 66 next to the waggon way).

2nd September 1799. To Cuthbert Forster, baker: a piece of ground from north to south 10 yards and from east to west 7 yards with a tenement or dwelling house and bakehouse boundering upon premises now in the occupation of Thomas Gleghorn on the east upon certain other premises now in the occupation of Mrs. Robson on the west upon the Waggon Way on the north and upon lands and grounds belonging to Sir Matthew White Ridley on the south.

(Note:—This site appears to be in the top corner of field No. 65 next to the waggon way.)

12 December 1800. To George Lough: all a piece or parcel of ground situate and being at South Blyth aforesaid and near to the Low Quay There containing in length from east to west fifty three feet or thereabouts and in breadth from north to south eighteen feet or thereabouts and also that messuage or tenement now used as a butchers shop and stable and office made erected and built upon the said piece or parcel of ground by him the said George Lough and now in his own occupation together with a fold yard for cattle situate between the said shop and stable boundering upon a raff yard and premises belonging to the said Sir Matthew White Ridley and now in the possession or occupation of Messieurs John Marshall and Mark Marshall on or towards the south and upon certain waste ground and premises belonging to the said Sir Matthew White Ridley on

83

or towards the east or west and north or by other the right meets and bounds and also all that Dung Hill on the west side of the said stable and adjoining thereto for the use of the said butchers shop containing in length from east to west ten feet and in breadth from north to south twelve feet or thereabouts.

Rent, 5s. For 39 years.

38. *The building of the road from Seaton Sluice to Blyth, 1795.*

The following extracts are taken from a series of letters written to Lord Delaval by John Bryers, his agent at Seaton Delaval.

April 20th 1795

. . . . The road your Lordship mentions is made as far as the Mile Hill, within about five or six yards of Mr. Ridley's new Stone fence all the way from the Wall corner below the Fountain Head and is to be continued to Blyth chiefly for the purpose of cannon etc. being readier conveyed if occasion was to be for them, & I understand particularly ordered to be made by General Balfour for this purpose I also hear that the Justices at some of their meetings in consequence of the necessity that appeared for a Hard good road to be made betwixt this place and Blyth ordered three hundred pounds to be laid out in making it Mr. Hannay of Blyth being the Chief manager and Mr. Crooks and Mr. Ridley I understand has also had directions about it, they make it by levelling down the Sand Hills where they happen in the way lays Stones under and Glasshouse rubbish above which seems to carry the Carts etc. very well; Carts are Hired for the purpose, but its supposed the money will not be sufficient to complete the whole. Mr. Ridley being surveyor of the roads is to pay them for this work. . . . .

May 6th 1795

. . . . The Military road along the Links goes but Slowly forward, few carts can be got to come for the wages of five shillings per day, and I understand Mr. Fenwick etc. has ordered Mr. Ridley to call on the Tenants (in this Manor) to do their Statute work there this summer if such hired carts cannot be procured. . . . .

June 12th 1795

. . . . They have begun to drive piles for a Cart Bridge to be made of Timber over Meggy's Burn, but the military road gets slowly forward towards it, being considerably short of Link House . . .

39. Details of Blyth trades and occupations extracted from the Earsdon Parish burial registers, 1763-99.

George Carmichael, Scoolteacher.	19 June	1763
Sarah, daughter of William Slack, Officer of Excise. ..	31 June	1764
John Martin, Officer of Customs.	2 Sept	1764
Thomas Killden, Sailor	10 June	1776
John Thompson, Labourer.	31 Dec	1776
Edmund Gibson, Servant to Mr. Hannay.	6 May	1777
Elizabeth, daughter of Timothy Duxfield, Farmer of Newsham.	1 June	1777
James Nicholson, Pilot.	31 July	1777
Thomas, son of Joseph Duxfield, Farmer of Newsham.	15 Dec	1778
Richard Wheatley, Master Mariner.	15 July	1779
Elizabeth, daughter of John & Catherine Elder, Roper.	29 Sept	,,
Richard Ord, Keelman.	9 Oct	,,
Ann, daughter of James & Isabella Thrift, Keelman.	23 Dec	,,
Sarah, daughter of Robert Smith, Shoemaker. ..	19 Jan.	1780
Mark Weatherhead, Mason.	30 March	,,
Thomas, son of Cuthbert Horsley of Linkhouse, Labourer.	29 June	,,
Jane, daughter of John Steel, Labourer.	9 July	,,
Elizabeth, wife of Nathaniel Cowell, Labourer. ..	30 Sept	,,
John, son of John Watts, Shipmaster.	20 Dec	,,
George, son of John Armstrong of South Linkhouse, Mariner.	27 Dec	,,
Lieut. George Mennell, Impress Officer.	28 Dec	,,
Cuthbert Hudson, Cordwainer (Shoemaker).	26 Jan	1781
Mary, wife of Edmund Hannay, Shipbuilder.	12 March	,,
George Richardson, Sailor.	24 April	,,
Jeremiah Nicholson, Sailor.	24 April	,,
William, son of John Elliot, Joiner.	29 April	,,
Edward, son of Edward Watts, Shipmaster.	2 Aug	,,
Margaret, widow of George Henry, Labourer. ..	30 Sept	,,
Elizabeth, daughter of Robert Stoker, Shipmaster. ..	5 Oct	,,
Thomas Hamilton, Sailor.	31 Jan	1782
Robert, son of Robert Jackson, Collector.	5 Feb	,,
Isabella, wife of Samuel Atkinson, Custom House Officer.	13 Feb	,,
Nathaniel Corby, Labourer.	17 Feb	,,
Ann, daughter of William Marchant, Shipmaster. ..	22 April	,,

Thomas Topliff, Sailor	14 May	1782
Robert Stoker, Sailor.	7 June	,,
Timothy Duxfield, Farmer of Newsham.	19 June	,,
George, son of Rev. John Thomas, Clerk	21 Sept	,,
Joseph Ingo, Officer of the Customs.	27 Sept	,,
Matthew Swinburn, Ships Carpenter	22 Dec	,,
Margaret, daughter of Edward Huitson, Sailor ..	27 Dec	,,
Elizabeth, daughter of George Shanks, Barber. ..	26 Feb	1783
William Wilkinson, Sailor.	19 June	,,
Eleanor, daughter of Thomas Mutter, Labourer. ..	19 June	,,
James, son of William Blakey, Taylor	19 June	,,
Elizabeth, wife of William Hague, Officer of Customs.	17 July	,,
Jane, daughter of Cuthbert Blakey, Taylor.	19 July	,,
Aaron Stoker, Sailor	31 Aug	,,
James Alexander, Labourer.	9 Sept	,,
John Thrift, Sailor	24 Sept	,,
Edmund, son of George Huntley, Butcher	4 Oct	,,
William Storey, Sailor	6 Oct	,,
Margaret Grozier, Salter.	3 Nov	,,
Aaron, son of Benjamin Corby, Sailor	22 Nov	,,
William Humble, Sailor.	13 Dec	,,
Witherington Young, Barber.	16 Dec	,,
Elizabeth, daughter of Philip Bradford, Sailor. ..	5 Jan	1784
John, son of Thomas Holmes, Labourer.	6 Jan	,,
Mary, wife of John Miller, Sailor.	23 May	,,
James, son of William Byers, Sailor	24 June	,,
John Nurse, Pilot...	6 July	,,
Ruth, Wife of John Storey, Butcher.	18 Aug	,,
John Hunter, Labourer.	27 Sept	,,
Robert Dobson of Link House Farm, Farmer. ..	4 Dec	,,
John, son of William Weatherhead, Mason.	24 Dec	,,
John Dale, John Pattison, James Whalie, William Harlem, William Donkin, Henry Taylor, Thos. Knox, Robert Rogerson, and William Poole, Sailors, drowned near Blyth and buried at Blyth Chapel ..	23 Jan	1785
James Phillips, Shoemaker.	24 Jan	,,
Thomas Atkinson, Labourer.	2 March	,,
Joseph, son of Robert Moor, Glazier.	18 March	,,
Ann, wife of Wm. Wilson of Newsham, Farmer. ..	19 April	1785
William son of William Lawry, a Traveller.	25 May	,,
John, son of Thomas Marshall, Sailor	20 July	,,
James, son of James Ogle, Shipmaster.	1 Aug	,,

John Wilson, Surveyor of the Customs.	16	Sept	1785
Jane, daughter of Joseph Nicholson, Labourer. ..	3	Nov	,,
Eleanor, daughter of William Weatherhead, Mason ..	22	Nov	,,
Thos. Farley, Joiner.	25	Nov	,,
Mary, daughter of Edward Watt, Shipmaster... ..	25	Nov	,,
Elizabeth, daughter of William Wilson, Officer of Excise.	15	Dec	,,
Mary, daughter of John Watt, Shipmaster.	25	Dec	,,
John Bruce, Sailmaker.	29	Dec	,,
Jane, daughter of Cuthbert Young, Carpenter ..	1	Jan	1786
Elizabeth, daughter of Cuthbert Young, Carpenter. ..	3	Jan	,,
Margaret Dawson, Spinster and Salter.	15	Jan	,,
John, son of William Miller, Shipwright.	13	Feb	,,
Ann, daughter of John Campbell, Sailor.	15	Feb	,,
Eleanor, wife of Thomas Orwin, Labourer.	1	March	,,
George, son of William Collier, Shipmaster. ..	11	March	,,
Jane, daughter of John Swinburn, Shipmaster. ..	14	March	,,
James, son of Richard Brown, Labourer.	20	March	,,
Jane, daughter of Ogle, Shipmaster.	28	March	,,
Thomas Wilkinson of Cowpen, Gentleman.	9	April	,,
Margaret, daughter of Gilbert Newton, Blockmaker.	18	April	,,
Hannah, wife of Jacob Wilson, Coastwaiter.	26	April	,,
Elizabeth, wife of Francis Francis, Ship's Carpenter.	3	May	,,
George Nicholson, Sailor	11	May	,,
Thomas, son of Matthew Linsley, Labourer.	12	May	,,
Mary, daughter of Joseph Readhead of Newsham, Labourer.	26	May	,,
William Wright, Officer of Customs.	25	June	,,
James Henderson, Joiner.	23	July	,,
Thomas, son of Thomas Young, Baker.	18	Aug	,,
John, son of Robert Moor, Glazier.	14	Sept	,,
Mary, daughter of Thomas Young, Baker.	23	Sept	,,
Thomas Reed, Labourer...	23	Oct	,,
Richard, son of Richard Gardener Charlton, Sailor	6	Dec	,,
Eleanor, daughter of Thomas Muter, Labourer. ..	13	Dec	1786
Hannah, wife of Adam Robinson, Sailor.	3	Feb	1787
James Smoult, Blacksmith.	20	Feb	,,
Elizabeth, twin daughter of Joseph Scaife, Agent. ..	21	March	,,
Ann, wife of Francis Wright, Master Mariner. ..	24	April	,,
Frances, daughter of Joseph Redhead of Newsham, Labourer.	9	May	,,
William Stewart, Gardener.	26	May	,,
John, son of Thomas Nesbitt, Roper.	31	May	,,

James Thrift, Labourer.	8 July	1787
Robert, son of Roger Ealson, Labourer.	11 July	,,
William Enos. a Traveller.	16 Aug	,,
Margaret, wife of Thomas Muter, Labouer, ...	4 Dec	,,
Eleanor, wife of John Swinburn, Shipwright. ..	18 Dec	,,
Francis Wright, Master Mariner.	21 Jan	1788
Matthew Newton, Sailor.	21 Feb	,,
John Watson, Labourer.	14 March	,,
Mary, wife of John Elliot, Ship's Carpenter.	24 April	,,
John Lidster. Blacksmith.	27 April	,,
Jonas, son of Jonas Polwart, Joiner.	8 May	,,
George, son of Robert Mitchell, Sailor.	5 July	,,
William Carr, Blacksmith	18 July	,,
Nicholas Wilson, Sailor	14 Sept	,,
Margaret, daughter of William Stoker, Sailor. ..	17 Sept	,,
Isabella, wife of Joseph Todd, Labourer.	23 Dec	,,
Barbara, daughter of Joseph Todd, Labourer. ..	23 Dec	,,
Barbara, daughter of William Watts, Shipmaster. ..	18 Feb	1789
John Watts, Shipmaster and Owner.	2 March	,,
Sarah, wife of George Detchin, Weaver.	4 March	,,
Margaret, wife of Thomas Wright, Labourer. ..	27 March	,,
William Hague, Tidewaiter.	7 May	,,
Edward, son of Matthew Robinson, Labourer. ..	14 June	,,
John, son of Thomas Nesbitt, Roper.	21 June	,,
Robert, son of William Lamb of Linkhouse, Gardener.	4 Aug	,,
Margaret, wife of Robert Mitchell, Sailor.	12 Sept.	,,
William, son of James Mattison, Brewer. ..	24 Oct	,,
Margaret, daughter of Thomas English, Surgeon. ..	1 Jan	1790
William, son of William Oliver, Blacksmith. ..	6 Jan	,,
William, son of William Pollard, Sailor	10 Jan	,,
Charles, son of Thomas Huntley, Butcher	20 Jan	,,
Sarah, daughter of Edward Watts, Shipowner. ..	21 March	,,
George, son of William Carr, Blacksmith.	15 May	,,
Adam Furguson, Shoemaker.	30 May	,,
Margaret, wife of John Camell, Sailor.	6 June	,,
Margaret, daughter of John Campbell, Sailor. ..	21 Aug	,,
John, son John Reed, Labourer.	12 Sept	,,
Margaret, daughter of Thomas. Freeman, Shoemaker.	17 Sept	,,
William Young, Sailor.	7 Oct	,,
Mary, daughter of John Milburn, Agent.	29 Oct	,,
Thomas, son of John Hopper, Sailor.	1 Nov	,,
William Hannay, Merchant.	9 Jan	1791

Thomas, son of Edward Atkinson, Sailor	22 Jan	1791
John Cock of Linkhouse, Malt Maker.	8 Feb	,,
Josiah, son of William Carr, Blacksmith.	5 March	,,
Ann, daughter of William Watts, Pilot.	6 March	,,
James, son of Henry Wilson, Farmer of Newsham.	18 April	,,
John Paton of Linkhouse, Maltman.	22 April	,,
John Hannay, Gentleman.	6 May	,,
Richard Gibson, House Carpenter	24 May	,,
Hannah, daughter of Martin Gleghorn, Sailor	31 May	,,
Thomas Alder, Shipwright.	8 June	,,
Jane, wife of William Carr, Blacksmith.	14 June	,,
Thomas Arkle, Labourer.	17 July	,,
Margaret, daughter of Thomas Foreman, Joiner.	21 Aug	,,
Joseph Ramsay, Shipmaster.	13 Sept	,,
Edward Potts, Labourer	2 Nov	,,
Margaret, daughter of Thomas Cooper, Labourer.	22 Feb	1792
Margaret, wife of John Balks, Sailor.	20 March	,,
John, son of John Milburn, Agent.	18 June	1792
Joseph, son of Joseph Redhead of Newsham, Labourer.	28 Aug	,,
William Oswell, Officer of Customs.	15 Nov	,,
Robert, son of Robert Bambray, Labourer.	18 Nov	,,
Nicholas, son of Nicholas Crummay, Sailor	5 Jan	1793
Jonas Polwart, Cabinet Maker.	28 Feb	,,
Elizabeth, wife of William Humble, Sailor.	10 March	,,
Ann, daughter of William Humble, Sailor.	19 March	,,
John Elliot, Joiner.	5 April	,,
John Morrison, Cordwainer.	7 May	,,
Thomas Gibson, Shipowner.	25 May	,,
George, son of Robert Brown, Blacksmith.	19 June	,,
John Hunter, Sailer.	21 June	,,
William Turner, Labourer.	28 Aug	,,
John, son of William Ridley, Labourer.	8 Sept	,,
Elizabeth, wife of Thomas Wilson, Shipmaster.	28 Feb	1794
Margaret, daughter of James Gogg, Labourer.	24 Feb	,,
Catherine, wife of William Stoker, Sailor.	26 Feb	,,
Sarah, daughter of Richard & Sarah Stoker, Trimmer.	2 March	,,
Margaret, daughter of John & Jane Richardson, Sailor.	18 Apl	,,
George Brown, Blacksmith.	15 June	,,
Andrew, twin son of Alex. & Mary Nesbitt, Roper.	20 July	,,
James, son of William & Jane Turner, Brickmaker.	1 Sept	,,
Jane, daughter of John & Mary Alder, Shipwright.	11 Sept	,,
Thomas Wright, Brickmaker.	13 Sept	,,

Joseph Hopper, Sailor.	9 Nov	1794
John, son of Shotton & Jane Smoult, Sailor. ..	16 Nov	,,
Thomas Forster, Sailor.	24 Nov	,,
Robert, son of Robert & Frances Gorby, Sailor. ..	2 Dec	,,
Richard, son of Cuthbert & Jane Blakey, Taylor. ..	26 Jan	1795
Ann, daughter of John & Rosanna Parker, Sailor. ..	3 Feb	,,
Mary, daughter of William Humble, Sailor.	8 Feb	,,
John Swinburn, Carpenter.	2 March	,,
Mary, wife of Robert Mafflin, Sailor.	22 March	,,
Ann, daughter of William & Jane Watt, Pilot. ..	29 March	,,
Jane, wife of James Lowther, Fiddler.	16 May	,,
William Harrison, Shipowner.	19 Sept	,,
John Boyd, Labourer.	20 Nov	,,
James, son of Henry Taylor, Sailor	10 Dec	,,
David Smith, Taylor.	7 Feb	1796
Jane, wife of William Bates, Sailor.	11 Feb	,,
Elizabeth, daughter of James Hogg, Labourer. ..	17 Feb	,,
Robert, son of John Milburn, Agent to Sir M. W. Ridley.	10 March	,,
Jane, wife of Shotton Smoult, Sailor.	11 April	,,
Ann, wife of John Robinson, Labourer.	14 May	,,
James Wilson, Sailor.	17 May	,,
Mary, daughter of William Ridley, Ropemaker. ..	15 June	,,
Jane, wife of George Morrison, Salt Officer. ..	10 July	,,
Timothy, son of Timothy Redhead, Labourer of Newsham.	31 July	,,
George, son of Edward Wright, Shipmaster.	22 Nov	,,
Joseph Turner, Brickmaker.	18 Dec	,,
James, son of Adam Cooper, Sailor.	15 Jan	1797
Sarah, wife of William Fenwick, Joiner.	23 Jan	,,
William Brown, Sailor	22 Feb	,,
Robert, son of Edward Watts, Master Mariner. ..	5 April	,,
Cuthbert, son of Cuthbert & Ann Gibson, Sailor. ..	8 April	,,
Mary, daughter of Edward & Mary Atkinson, Sailor.	9 April	,,
James, son of James & Alice Mills, Mason.	20 April	,,
Henry Elder, Sailor.	8 May	,,
Sarah, daughter of William Fenwick, Joiner. ..	7 May	,,
Edward Nicholson, Sailor.	19 May	,,
Thomas Swinburn, Sailor.	25 May	,,
John, son of John Turnbull, Sailor.	15 June	,,
Hannah, daughter of James Davidson, Sailor. ..	27 June	,,
Barbara, wife of William Wigham, Sailor.	31 July	,,

John, son of George Storey, Butcher.	14 Aug	1797
Ann, daughter of William Stephenson, Keelman. ..	19 Sept	,,
Mary, daughter of Jeremiah Hunt, Keelman. ..	13 Nov	,,
Mary Manners, widow of the late John Manners, Sailmaker.	9 Dec	,,
Martha, wife of Robert Shanks, Barber.	3 April	1798
James Walker, son of John Walker, Sailor.	14 July	,,
Richard, son of Thomas Birkley, Brewer.	5 Aug	,,
Joseph Ramsay, Gentleman.	17 Aug	,,
Ann, daughter of Henry Patton, Sailor.	16 Oct	,,
Robert Brown, Blacksmith.	17 Nov	,,
Catherine, daughter of James Davison, Agent. ..	18 Nov	,,
William, son of John Clark of Crofton, Roper. ..	12 Dec	,,
John Lamb, Carpenter.	18 Jan	1799
Isabella, wife of William Gowland, Watchmaker. ..	14 Feb	,,
Elizabeth, daughter of William Warner, Sailor. ..	20 Feb	,,
Andrew, son of Thomas Freeman, Shoemaker. ..	20 Feb	,,
Elizabeth, daughter of Thomas Freeman, Shoemaker.	21 Feb	,,
Margaret, daughter of John Brockett, Carpenter. ..	3 April	,,
Margaret, wife of John Charlton, Grocer.	8 April	,,
Elizabeth, daughter of John Morrison, Sailmaker. ..	21 May	,,
John Stoker, Sailor.	30 June	,,
Elizabeth, daughter of William Redhead, Ropemaker.	13 Aug	,,
Robert Mafflin, Sailor.	27 Sept	,,
Thomas, son of William English, Surgeon.	11 Nov	,,
Jane, widow of George Marshall, Timber Merchant.	20 Nov	,,

40. *Summary of Blyth trades and occupations, 1763-99.*

Sailors	.. 75	Tailors	.. 4	Blockmaker	.. 1
Labourers	.. 38	Butchers	.. 4	Clerk ..	.. 1
Shipmasters	.. 19	Brickmakers	.. 3	Collector..	.. 1
Carpenter, Joiner	15	Masons	.. 3	Fiddler ..	.. 1
Ship's Carpenter	7	Sailmakers	.. 3	Grocer ..	.. 1
Blacksmiths	.. 11	Barbers	.. 3	Impress Officer	1
H.M. Customs	9	Salters	.. 3	Manservant	.. 1
Shoemakers	.. 8	Glaziers	.. 2	Merchant	.. 1
Ropers	.. 7	Excisemen	.. 2	Salt Officer	.. 1
Agents..	.. 5	Gentlemen	.. 2	Schoolmaster	.. 1
Shipowners	.. 4	Bakers	.. 2	Surgeon	.. 1
Farmers	4	Gardeners	.. 2	Timber Merchant	1
Keelmen	.. 4	Brewers	.. 2	Trimmer	.. 1
Pilots ..	.. 4	Maltmakers	.. 2	Watchmaker	.. 1
				Weaver	.. 1

Bibliography

PRINTED

History of Blyth (1862), by John Wallace. This pioneer work contains much usefull material but is far from being a complete account of the town's history.

The History of the Port of Blyth (1929), by C. A. Baldwin, who for many years was the chief officer of the port.

The Story of Blyth (1957), published by *The Blyth News and Ashington Post*. Although this popular account of the town relies heavily on the earlier histories it also contains additional information collected by the local newspaper's own editorial staff.

A History of Northumberland, (Volume IX): *The Parochial Chapelries of Earsdon and Horton* (1909), by H. H. E. Craster.

The Newcastle Journal, from 1749.

DOCUMENTARY

The Ridley of Blagdon MSS. The wealth of evidence - in account books, letter books, leases, maps, estate papers and correspondence - concerning the administration of a large estate comprising varied agricultural, in-dustrial and commercial interests makes these family archives one of the most important sources for Northumberland's economic history.
Earsdon Parish Registers.
Records of registration of ships with the Customs and Excise Department, held at various east coast ports.

LIST OF REFERENCES TO DOCUMENTS

The documents referred to below by the prefix *ZRI* are among the *Ridley of Blagdon MSS.* deposited in the Northumberland Record Office.

Frontispiece View of Blyth by Balmer, about 1827, from original in possession of Blyth Corporation.

1. *ZRI.7.*
2. *ZRI. 37/2.*
3. *ZRI. 49/4* The map numbers in the first column are the field reference numbers on the plan of the township of Newsham and South Blyth, 1840 (44. below). The rents are endorsed on the document itself.
4. *ZRI. 49/5.*
5. *ZRI. 49/9.*
6. *ZRI/37/3/4.*
7. *ZRI. 37/3/2.* The production is also stated fortnightly.
8. *ZRI. 37/3/2.*
9. *ZRI 37/3/2.* Cross references to pages in the Journal are omitted.
10. *ZRI. 37/4/2. passim.* These are only a few select entries of special interest. Cross references to other account books are omitted.
11. *ZRI. 37/3/2.* Cross references to entries in the Journal are omitted.
12. *The History of the Port of Blyth*, Baldwin, contains the full text of Rennie's report, of which this is an extract only.
13. *ZRI. 37/4/1.* The figures for Jan-April 1793 are taken from ZRI. 37/5.
14. *ZRI. 37/4/1, 6.*
15. *ZRI. 37/4/1, 6.*
16. *ZRI. 37/3/2.*
17. *ZRI. 38/2.*
18. *ZRI. 35/25.*
19. *ZRI. 35/25.*
20. *ZRI. 35/25.* This document is undated but the paper is watermarked "GILLING & ALLFORD 1822".
21. *ZRI. 35/25.*
22. *ZRI. 37/3/2.* Cross references to other account books are omitted.
23. *ZRI. 37/4/2.* Cross references to other account books are omitted.
24. *ZRI. 37/4/2 37/3/3, 4.*
25. *ZRI. 38/2.* The letter from the Duke of Northumberland is from ZRI. 25/6.
26. *ZRI. 37/3/2.*

27. *ZRI. 37/3/2.*
28. *ZRI. 37/3/2.* Select entries only. Cross references to other accounts omitted.
29. *ZRI. 37/3/3.*
30. Various old tables of duties.
31. *ZRI. 4/4.*
32. *ZRI. 37/4/2.* As 28.
33, Extracted or compiled from records of registration of ships in the
35. Customs and Excise Dept., and Lloyds Register.
36. *ZRI. 43/1.*
37. *ZRI. 4/4.*
38. *2 DE. 4. 22/1-77.*
39, Earsdon parish burial registers in NRO. This list includes only
40. entries which specify the occupation of the deceased, or in the case of infants the father's occupation.
41. *ZRI. 49/9.*
42. *ZRI. 37/4/1.*
43. *ZRI. 38/2.*
44. *ZGI. XXI/3.*

July 2ᵈ 1759

	Dr	Do	Cr
Tryal of Salifleet			
John Curtis Master			
8 Chaldrons of Coals at	4	5	
Harb'r Dues		1	4
	4	9	4	By his on Acct
Do 25			Note	4	9	4
John & Margery of Blyth			
Geo Huntley Master			
71 Chaldrons of Coals at	40	14	
Ballast & Harb'r Dues		13	
	41	7	By Edw Hannay
Do 25			Go Note	41	7
Francis of Lynn			
Peter Sorensen Master			
41¾ Chald'rs of Coals at	22	18	4
Ballast & Harb'r Dues		11	4
To Cash to him then	4	4	By his Bill on Messrs Sartees
	27	13	8	& Burdon	27	13	8
Do 26			
Speedwell of Lynn			
John Stevenson Master			
3⅓ Chald'rs of Coals at	1	16	8
Harb'r Dues		1	4
	1	18	By Cash Recd	1	18
Do 30			
Three Brothers of Wisbeach			
Rich'd Perry Master			
49½ Chald'rs of Coals at	27	0	4
Ballast & Harb'r Dues		12	
To Cash to him then	3	3	By his Bill on M'r Walter
Do Aug 1759	31	14	Wright	31	14
Mary of Blyth			
John Hannay Master			
4 Chaldrons of Coals at	1		
Harb'r Dues		1	4
	1	1	4	By Cash Recd £ 1 1 4

Waggon Way West Farm. Tim. & Jos. Dukinfield.

No.	Names of Fields.	Quantity in each Field.			Quantity in each Farm		
		A	R	P.	A	R	P.
	Brought forward				1101	1	37
1	West Waggon Way Field	28	0	28			
2	Housesteads	23	3	32			
3	House, Stack Yards & Waste	9	0	25			
4	East Waggon Way Field	29	0	18			
5	East Middle Field	24	1	16			
6	Laverick Hall Field	14	2	8			
7	West Middle Field	14	1	27			
	Lane in Do.	0	1	22			
8	Stickley West Field	28	3	8	165	3	20
	Carried forward				1267	1	17

GROUND. * COOPEN.

Waggon Way.

LORD DELAVAL'S

West Farm.

No.
ml.
1

2

3

4

5

6

7

8

Newsham West Farm

No. 11.

WAGGON WAY

WEST FARM

— part of —

Newsham

Lordship.

Newsham West Farm.

Scale of Chains.

Newsham West Farm. Jno. Dukesfield Jur.

No	Names of Fields.	Quantity in each Field			Quantity in each Farm		
		A	R	P	A	R	P
	Brought forward				1267	1	17
9	Newsham West Field	27	3	19			
10	Houses, Stack Yards &c	1	2	25			
11	Little Field	5	3	32			
12	Well Close	15	0	8			
13	Camp Field	19	3	30			
14	West Blakeburn	12	2	5			
15	Borehole Field	13	0	33			
16	Middle Blakeburn	10	2	21			
17	Bawseys Close	11	3	10			
18	Great Field	18	3	2			
19	East Burn Close	10	2	12	148	0	1
	Carried forward				1415	1	18

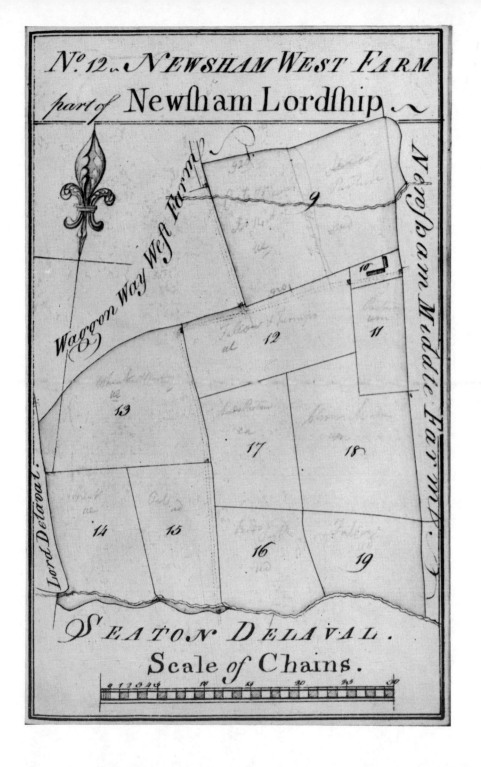

Newsham Middle Farm. Jas. Clarke

Nº	Names of Fields.	Quantity in each Field			Quantity in each Farm		
		A.	R	P	A.	R	P
	Brought forward				1415	1	18
20	West Blakeburn	16	3	4			
21	East Blakeburn	13	2	22			
	Dº Bog	2	0	22			
22	Middle Field	18	2	20			
23	Shoulder of Mutton	16	1	8			
24	West Field	17	2	4			
25	East Field	11	1	14			
26	House, Yards, Garden & Waste	3	2	8			
27	West Greens	14	2	15	114	1	37
	Carried Forward				1529	3	15

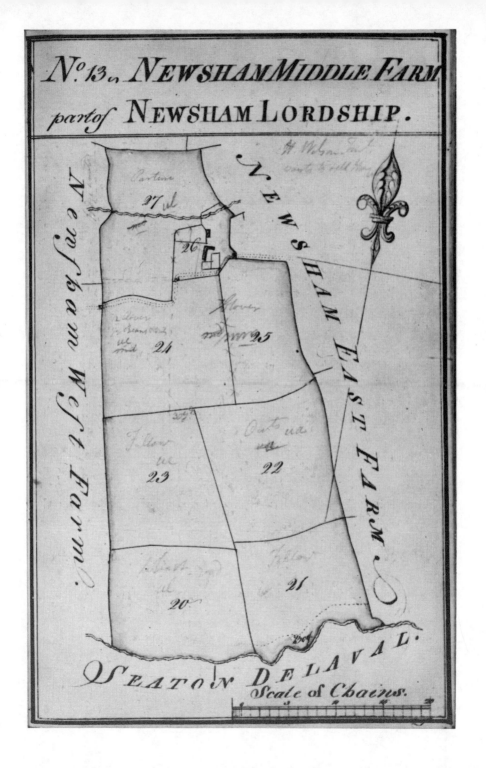

Nº 13, NEWSHAM MIDDLE FARM

part of NEWSHAM LORDSHIP.

SEATON DELAVAL.

Scale of Chains.

Newsham East Farm. Wm Bennett.

N°	Names of Fields	Quantity in each Field			Quantity in each Farm		
		A	R	P	A	R	P
	Broughts Forward				1529	3	15
28	House Stack Yard & Waste	3	3	2			
29	Bog Field	10	3	13			
30	Shoulder of Mutton	14	1	30			
31	Bog Field	12	1	2			
32	Broad Field	14	3	20			
33	Horse Pasture	10	1	26			
34	Long Close	17	0	6			
35	West Blakeburn	19	0	27			
37	East Blakeburn	10	3	21			
36	Middle Blakeburn	16	9	24	130	1	11
	Carried forward				1660	0	26

Waggon Way East Farm.

Pasture
no. 29

28

28

Newsham Middle Farm.

30

The End

31

32

33

34

35

36

37

The Link Farm.

Link House Farm.

Scale of Chains.

No. 14.

NEWSHAM
EAST FARM.

Nich: Ridley Esquire.

Seaton Delaval belonging to Sir. Delaval.

Waggon Way East Farm. John Watson.

Nº	Names of Fields.	Quantity in each Field A R P	Quantity in each Farm A R P
	Brought forward		1660 0 26
38	Horse Pasture	45 1 12	
39	Broad Flatt	16 1 15	
40	Newsham Field	22 3 26	
41	Roe Pasture	11 2 30	
42	The March	7 1 8	
43	Stack Yard, Garden & Waste	1 3 7	
44	Garden Field	16 3 14	
45	Broad Field	19 2 15	
47	Middle Field	9 1 35	
48	Link Field	9 3 26	131 0 28
	Mr Rich.d Jackson's Farm.		
63	Jackson's Close		7 0 2
	Carried forward		1798 1 16

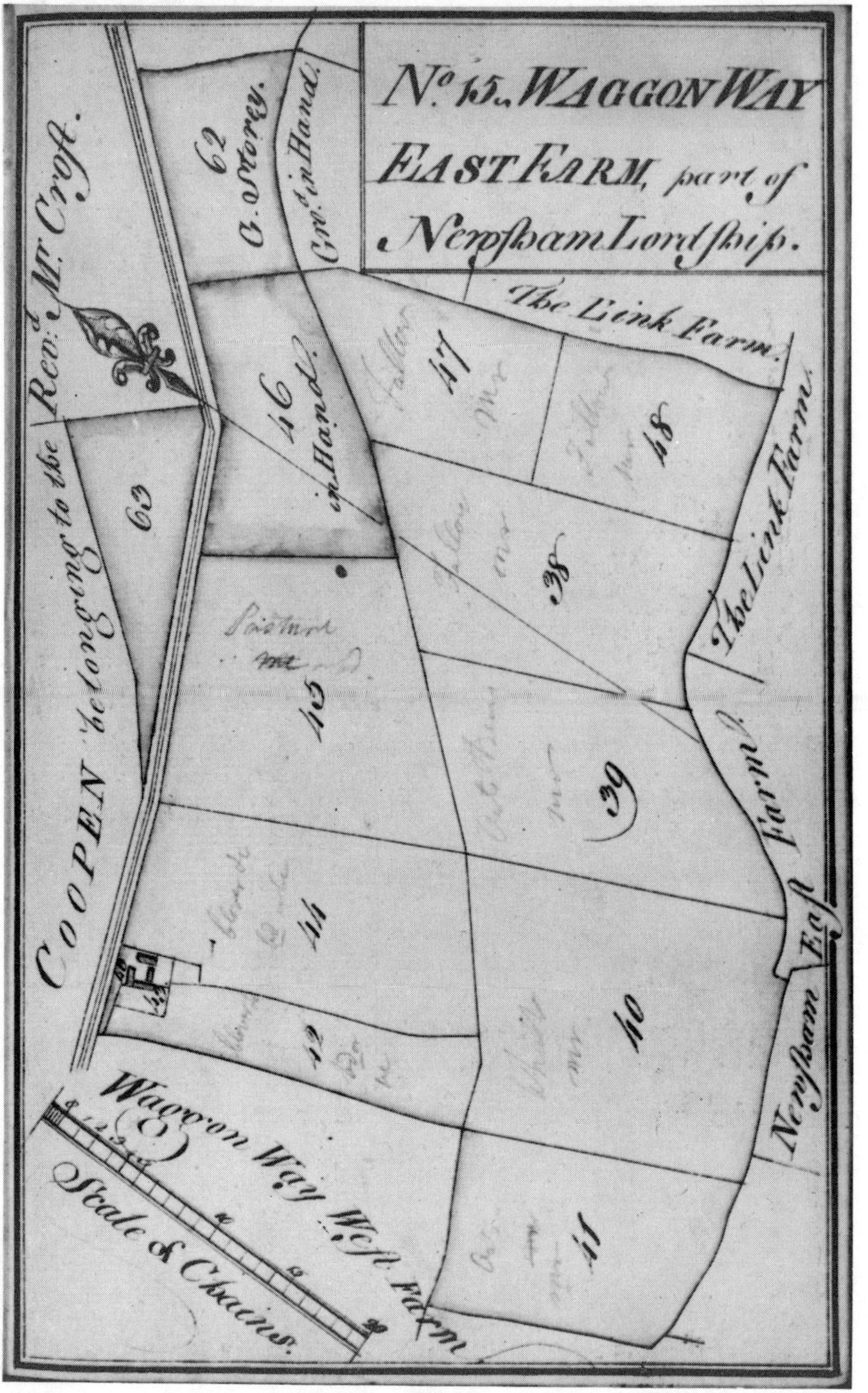

No. 15. WAGGON WAY EAST FARM, part of Newsham Lordship.

62 G. Storey.

Gov. in Hand.

The Link Farm.

46 in Hand.

Fallow 47

Fallow 48

38

63

COOPEN belonging to the Revd. Mr. Croft.

Picture 45

39

The Link Farm.

44

43

Newsham East Farm.

42

40

11

Waggon Way West Farm.

Scale 8 Chains.

Link Farm — Elstob & Hoog.

Nᵒ	Names of Fields	Quantity in each Field			Quantity in each Farm		
		A	R	P	A	R	P
	Brought forward				1898	1	16
49	Near Barley Close	22	3	2			
50	Far Barley Close	14	1	20			
51	Prior's Close	9	1	25			
52	South Link Close	15	3	1			
53	North Dᵒ (including y.ɡ.)	15	2	0			
54	High Field	20	1	8			
56	Sheet Field	18	1	28			
		116	2	0			
57	Clay Hole	5	0	0			
58	Link Pasture	33	1	6	154	3	6
	Joseph Turner's Farm						
45	Shotton's Close				11	2	23
	Carried forward				1961	3	5

No. 16.

LINK FARM,
part of
Newsham
Lordship.

Ropery.

Ropery.

58

57

56

55

inland

53

54

52

51

Waggon Way

G. Storey

John Storey

Newsham Road

N° K I L L E N

K S

First Fa

Fis.

Blyth. (Mrs Marshall's) Farm

Nº	Names of Fields	Quantity in each Field			Quantity in each Farm		
		A	R	P	A	R	P
	Brought forward				1961	3	5
59	Low Close				6	3	30
	John Storeys Farm						
60	Low Middle Close				11	0	34
	George Storeys Farm						
61	High Middle Close	7	0	32			
62	High Close bro. from Plan Nº 35	8	0	4	15	0	36
	Messrs Clark & Watts Farm						
63	South West Field	4	0	24			
64	South East Field	3	0	24			
66	North East Field	3	2	4			
67	North West Field	3	0	20	13	3	32
	Grounds in Hand						
68	Waggonway Field	6	3	27			
69	Ropery Field	7	2	15			
70	Bean Field bro. from Plan Nº	11	3	24	26	1	26
	Wastes &c at Newsham						
	The Sinks	54	2	17			
	Ballast Hills	5	2	26			
	Waggonway	10	3	34			
	Blyth Town	16	0	4	87	1	1
	Carried Forward				2125	3	4

Scale of Chains.

BLYTH RIVER.

Salt Pans.

Ballast Hills.

Stand Dry at Low Water.

COOPEN.

66

67

68

65

64

69

N.º 17.

BLYTH FARMS.

part of Newsham

Lordship.

The Link Farm.

59

COOPEN.

60.

61

Link House Farm, belonging to Nicholas Ridley Esq.r & Harry Dobson Ten.t

No	Names of Fields.	Quantity in each Field			Quantity in each Farm.		
		A	R	P	A	R	P
	Brought forward				2125	3	1
70	Little Close	6	1	26			
71	Back of Garden	1	0	32	7	2	18
72	Broad Pasture	23	3	20			
73	Barn Close	9	0	20	9	0	20
74	Link House Close	13	3	26	13	3	26
75	Newsham Field	10	3	20	10	3	20
76	West Clover Field	12	3	22			
77	East Clover Field	13	3	7			
78	Link Close	11	0	18			
	Plantation in D.o	0	1	25			
79	South Intake	5	2	14			
		108	3	30			
80	Burnside	5	1	26			
81	South Links	23	0	35			
82	North Links	16	2	24			
83	Garden &c	2	1	10	156	1	1
					2282	1	5

No. 18. LINK HOUSE FARM.

part of Nerosham Lordship, belonging to N. Ridley Esq?

Link Farm.

The Link Farm.

Pasture
72

70

71

82

83

Beans
73

Link House.

Wheat ½ Oats
74

Clover
75

½ Wheat ½ Oats
74

½ Oat ½ Seed
78

Fallow
76

Seeds
77

81

Fulwood
79

Nerosham East Farm

Nerosham E. Farm.

80

The Sands Dry at Low Water.

Scale of Chains.

Geo. Wards Esq.   Newcastle 13 Septr 1776

I see by your Favour of the 9 that the Coal Markett is very gay
as to prices. I have not any thing to advise you of from Blyth, but
that the several Ships mentioned to you in my last besen for London
are Sailed except Briggs. I am sure that there can be no mistakes
to the Mixture of round & small there, but it is possible that some
Ships cleared for the Coast have made a Trip to London which Jois
may be enquired into. I never saw rounder Coals than stied on
Wednesday last. I am

                                        Yours &c M. Ridley

Geo. Wards Esqr   Newcastle Septr 20. 1776

I am glad to see by your Favour of the 16 that the Blyth Ships
are arrived at London & sold well. you may depend, that those Coals will
be kept up to their Roundness; there is not one Ship in that
Harbour for London, Briggs being Sailed; I hope we shall soon
have a Supply   I am

                                        Yours &c.
                                        Mr Ridley

Geo. Wards Esqr   Newcastle September 27. 1776

I am favoured with yours of the 23 not any Ships have sailed from
Blyth for London since my last. Rich. Wheatly & Jas. Brown are in the
Harbour and will be dispatched for the same as soon as possible I will
deliver at Blyth any Quantity of Salt even 500 Tons, as the Garners are full &
the same laid off; the present price is 26/ p Ton. but if that price cannot
be had I will take 25; I shall be much Obliged to you if you can get a
Quantity disposed of.   I am yours   M. Ridley

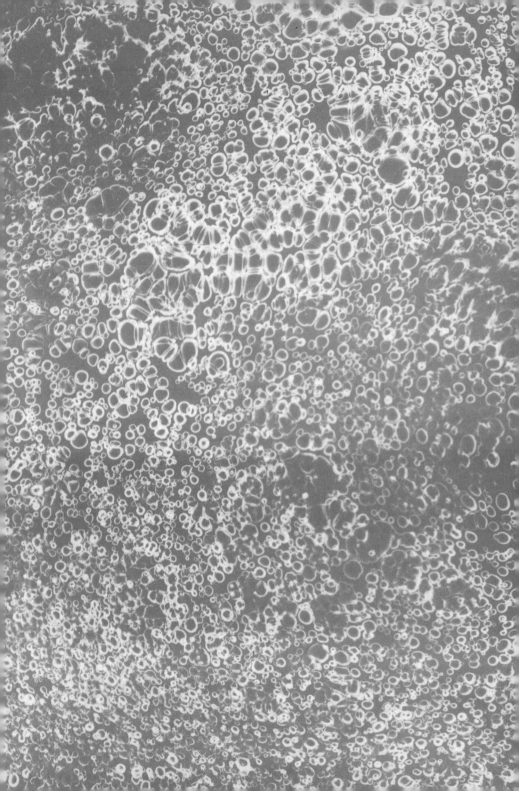